ID0982662

The Aftermath of Defeat

The Aftermath
of Defeat

Societies, Armed Forces, and the
Challenge of Recovery

Edited by
George J. Andreopoulos and Harold E. Selesky

. . . .

Yale University Press   New Haven and London

Published with the assistance of the Frederick W. Hilles Publications Fund of Yale University.

Designed by James J. Johnson and set in Ehrhardt type by Tseng Information Systems, Inc., Durham, North Carolina.
Printed in the United States of America by BookCrafters, Inc., Chelsea, Michigan.

Library of Congress Cataloging-in-Publication Data

The Aftermath of defeat : societies, armed forces, and the challenge of recovery / edited by George J. Andreopoulos and Harold E. Selesky.
    p. cm.
    Includes bibliographical references and index.
    ISBN 0-300-05853-5

    1. Combat.   2. War and society.   3. Military policy.
    I. Andreopoulos, George J.   II. Selesky, Harold E.
    U21.2.A33   1994
    355.02'8—dc20   94-3989
                                                CIP

A catalogue record for this book is available from the British Library.
The paper in this book meets the guidelines for permanence and durability of the Committee on Production Guidelines for Book Longevity of the Council on Library Resources.

10 9 8 7 6 5 4 3 2 1

Contents

Preface

Most of the chapters in this book were first presented at Yale University under the auspices of the International Security Program. Every year these lectures, which are funded through the generous assistance of the Lynde and Harry Bradley Foundation, deal with a particular theme in military and strategic history. On this occasion, the speakers were asked to focus on the idea of recovery from defeat.

This volume could not have come into being without the aid of a number of people. In particular, we would like to thank our colleague Paul Kennedy, Dilworth Professor of History and director of the International Security Program, for his encouragement and support throughout this project. Ann Bitetti of the International Security Program patiently typed and retyped successive drafts of the manuscript; we owe her more than we can ever express in words. Marge Camera provided indispensable assistance in preparing this volume. At Yale University Press, it was a great pleasure to work with Charles Grench, executive editor, Otto Bohlmann, assistant editor, and Jane Hedges, senior manuscript editor; their strong interest in the project coupled with their patience and sound advice made our task much easier. A special note of thanks must go to Giuliana and Joyce for their continuing support and assistance.

Above all, our thanks go to the individual contributors. We sincerely hope that the final product meets with their approval.

GEORGE J. ANDREOPOULOS
HAROLD E. SELESKY

The Aftermath of Defeat

Introduction

. . . .

Assessing Recovery

George J. Andreopoulos and Harold E. Selesky

In recent years, the study of military and naval history has expanded far beyond the traditional emphasis on battlefield tactics and campaign strategies. While these subjects still form the focus of major scholarly studies, many military historians have come to realize that it is equally important to analyze and assess the societal matrix within which the more specifically military challenges materialize. It has become an axiom in military history that war has a profound impact on the society that engages in it, and that the more extensive the involvement, the greater the strain war-making exerts on that society. This strain can prove overwhelming to the losing side: defeat is the most shattering short-term experience any society can encounter. In many cases, however, the shock of defeat lingers in the national psyche with disturbing long-term implications for the society and its institutions, as, for example, when the Nazis skillfully exploited Germany's defeat in the First World War to undermine the fragile institutions of the Weimar Republic.

Examining the manner in which a society recovers from the trauma of defeat offers a way to delve more deeply into the reciprocal connections between the waging of war and the evolution of society, and a way to view both sides of the relationship in a new light. In particular, such a perspective can offer useful insights into how a certain society has perceived the challenges confronting it at a critical juncture of its history, as well as the extent to which the demands of recovery have had an impact on the reconsideration of fundamental institutional arrangements.

The aim of this book is to examine the proposition that defeat in war can have a profound impact that extends well beyond the battlefield—

an impact on society and its ability to redefine both its short- and long-term policy goals. The eight case studies were chosen to reflect a wide range of times and places as well as types of conflict (from the American Revolutionary War to China's invasion of Vietnam in 1979; military and naval, conventional and counterinsurgency warfare).[1] In addition, they reflect different perspectives on the link between the recovery of the armed forces and reform in the society at large. Some authors consider recovery and reform as inextricably linked, being part of an essentially political process. Others view the link between recovery and reform as tentative at best; their case studies demonstrate that "in-house" reforms confined within the services and the decision-making processes of the government (whether in the form of resource allocation, of containing interservice rivalries, or of reorganizing a whole department) achieved the desired results with very few repercussions for the society at large. The assessment of the long-term implications is also varied. In some cases, the need for domestic reforms led to the consolidation of the status quo; in others, it simply exposed the inability of the traditional social order to cope with the strains generated, and this led to disturbing outcomes; while in certain others it led to a showdown between the officer corps and the government and the reassertion of the long-overdue civilian control over the military.

All these studies share at least one characteristic, however. They all treat the issue of recovery on what might be called the level of grand strategy—that is, from a perspective that seeks comprehensively to integrate civilian and military resources at the highest reaches of a nation's ability to balance means and ends. Because the defeats about which these authors write were all-encompassing, their investigations of subsequent efforts to recover inevitably emphasize the fact that recovery is most often a process that proceeds from the top down. All recognize that, while some of the roots of defeat may lie in a failure to adequately understand changes in battlefield tactics, in an inability to incorporate technological innovations into tactical, operational, or strategic planning, or in the lack of foresight to formulate an appropriate strategy for a particular campaign, ultimate responsibility for success or failure rests on the capacity of the nations' top political and military leadership to bend all of their society's resources to master the challenges confronting them. Thus, regardless of whether defeat ultimately undermines the leaders' and the regime's legitimacy, it invariably generates pressures for some type of reform. In many instances, the leadership's viability hinges on its ability to accurately detect these pressures and harness them accordingly.

One major theme that emerges from these studies is that, due to the many variables involved, assessing recovery is an extremely complex issue. The type of evaluative criteria used to assess the process of recovery can vary not only in time and place but also in historical perspective; it is not uncommon for historians to radically reassess contemporary interpretations of what constituted success and failure.

At its most basic, the process of recovery is associated with a significant reversal of fortunes on the battlefield. Thus, it depends to some extent on what lessons the armed forces and the societies that sustain them are able and willing to learn from a major setback. Do the leaders of the armed forces admit their mistakes and are they willing to do what is necessary to achieve tactical and operational success? How successful are they in learning how to adapt and change under the intense pressure of a debacle during wartime? Armed forces may be judged to have recovered when their tactical and operational proficiency has reached a point that surpasses that of their opponent, when they have succeeded in demonstrating to their opponents that the tables have turned. This state of affairs might be achieved before the end of the conflict or might be seen in retrospect to have required several years or even decades of "peace"— a suspension of active hostilities—to reach full blossom.

The picture becomes more complicated once assessment moves beyond "progress" on the battlefield. If recovery is associated with the need for substantive reforms in the society at large, then not only does recovery become more difficult to assess, but the time frame within which the assessment can occur is often arbitrarily defined. Should a cycle of defeat and recovery be defined primarily on the basis of whether the society successfully deals with the next military challenge? And what if the next challenge is not of a military nature and yet the society under study overcomes it precisely because of the lessons learned from a military debacle? To what extent can military failures prepare societies to confront nonmilitary challenges? Moreover, to what extent can the successful handling of nonmilitary challenges signal recovery without a corresponding favorable military outcome? These questions raise a fundamental issue: the extent to which the defeat-recovery cycle can be conceptualized without elevating the military outcome as the overall determinant of the cycle's boundaries.

Two examples from this volume will suffice to highlight the complex facets of this issue: the Italian recovery after Caporetto and the French recovery after the Indochinese and Algerian debacles. In the first case, the humiliating experience of Caporetto was reversed a year later with the victory at the Battle of Vittorio Veneto, a victory that signaled the collapse of the Austro-Hungarian army. On the strictly military level, one could periodize the cycle of defeat-recovery around the battles of Caporetto–Vittorio Veneto. On another level, however, if recovery is to encompass societal pressure for reform, such a periodization appears misleading. When the role of social reform is considered, the victory at Vittorio Veneto does not represent the end of the defeat-recovery cycle but is viewed as an event which failed to reverse the declining fortunes of the established order. And similarly, the failure at Caporetto is not the beginning of the cycle but the manifestation of a deeper malaise—a malaise that, by unleashing powerful forces for reform, nurtured the potential for major postwar upheavals with the well-known disastrous consequences.

In the French case, while there are major military setbacks, no comparable victories exist; hence, the conceptualization of a militarily relevant defeat-recovery cycle is impossible. In the nonmilitary sphere, however, the military setbacks brought to the forefront the problem of France's colonial overextension and the need for a managed retrenchment—a retrenchment that would reflect France's shrinking influence in international politics.[2] This process of retrenchment rendered imperative the need to articulate a new mission for the nation's armed forces, a task brilliantly conceived and executed by de Gaulle. Although the French armed forces are yet to be "tested" in the battlefield, there is little doubt that their sustained focus on their professional mission and their unquestionable subordination to civilian rule are indicative of a long overdue recovery.

Thus, on the strictly military level, it is possible to conceptualize a cycle of defeat and recovery bounded by battlefield setbacks and victories. But once the focus shifts to the interaction between military objectives and society's ability to define and sustain these objectives, the picture becomes infinitely more complicated and resistant to facile generalizations. This book affords the reader much food for thought on how one might construct both a typology and a morphology of recovery from defeat. And although the evaluative criteria for such a typology may differ from case to case, the chapters in this book point in one direction: the most promising theoretical insights can be derived from the case study approach, rather than from the use of formal modeling techniques.[3]

These contributions suggest that one important analytical category is the time frame of recovery for both the armed forces and the society, especially whether or not recovery occurred in time for the society to profit from her exertions during the conflict. In the opening chapter, R. J. B. Knight shows how Britain's Royal Navy overcame a bad start in the War of American Independence. Lack of investment in naval power during peacetime, as well as during the first years of the American rebellion, left the navy in serious straits when France threw her naval and military might behind the American colonists in 1778. But an adequate resource base, the adoption of two major technological innovations (copper sheathing of ships' hulls and the carronade), and the foundations of a system of sound operational and administrative control allowed Britain to recover sufficiently, first to survive and then to turn the tables on her ancient foe. This may be construed as a case of overcoming problems of an operational nature, of having enough time for leaders to make a good system work. Even more important, although this could not have been known to the leaders at the time, the experience of having to contend with the French navy left the Royal Navy in a much better position when next it came time to hold the seas against France.

In the second chapter, Peter Paret looks at the broader issues associated with Prussian recovery after the defeat in Jena. He argues that it is impossible to understand the Prussian comeback without focusing on the nonmilitary elements and that the reform of the Prussian army as well as the reforms of the political structure and the economic practices of the Prussian society were parts of the same process. What makes the Prussian case particularly intriguing is that the pressure for long-overdue reforms emerged from a misreading of the causes of the 1806 defeat. It can be argued that although the Prussian army's organization, equipment, and training were not comparable to those of the Napoleonic army, the true causes of the defeat were inferiority in numbers and resources. The reformers' agenda (whether espousing significant changes in society as an end in itself or as a means to achieve a more radical military reform), coupled with the overextension of Napoleonic France, made possible the impressive comeback of the 1813–15 period. Once the Napoleonic threat faded, however, so did the pressure to complete the reform program; and the legacy of this stillborn reform process would come to haunt Prussian society and her institutions.

The third chapter by John Gooch explains how Britain responded to her failure to deal effectively with a colonial foe, the Boers in South Africa. As they had done at the start of the War of American Independence, the British at the end of the nineteenth century did not appreciate the difficulty of fighting a colonial war thousands of miles from home.

Perhaps because only Britain's reputation and prestige were at stake, not her national survival, the recovery period was limited to the army's efforts to undertake a thoroughgoing analysis of strategic, operational, and tactical errors, efforts that were both in accord with a faith in progress and the need to adapt to the accelerating pace of technological change. Institutional reorganization sought to remedy interservice rivalries, as well as failures in anticipating strategic, operational, and tactical needs, especially in areas such as intelligence. Here the dividends of recovery were seen most clearly in the next war, although Gooch does note the great irony that inaccurate conclusions drawn from observing Japanese forces in the Russo-Japanese War eroded what might otherwise have been an even more considerable recovery.

Brian R. Sullivan's chapter on the Italian recovery from Caporetto concludes the section that focuses on how societies can recover during wartime. As previously noted, the Italian army managed to reverse the humiliating experience at Caporetto with an impressive victory—within a year—at Vittorio Veneto. What contributed to this remarkable military comeback was a growing awareness among the Italian people that the conflict had transformed itself from yet another dynastic war into a national struggle against a foreign invader. The Caporetto debacle gave substance to a new sense of national uniformity; a uniformity that the Piedmont-based ruling elite had consistently failed to articulate in the past. This transformation in popular consciousness was sustained by the government's belated attempts to promote a new vision for postwar Italy; a vision promising a new economic order and inclusionary governmental policies. Herein lay the challenge and the danger: the ruling order had to deliver in order to survive. Caporetto had accentuated the manifold shortcomings of the old system and the war effort had temporarily consumed the energies generated by the exigencies of reform. But after the war, there were no safety valves available, and the ruling elites were soon confronted with their promises. Compared to the relative success of the Prussian elites in controlling and containing the explosive social forces unleashed by defeat, their Italian counterparts were found wanting. The rise of fascism represents one of the most stunning reversals in the fortunes of societal reform; coming on the aftermath of what was considered a remarkable military recovery, it was the supreme manifestation of how truly irrelevant the established order had become.

Like the first four chapters, the remaining four also form something of a group. All four—with some exceptions in the Israeli case—examine situations in which recovery can best be assessed by reference to capability and performance many years after the events which marked the defeat-recovery sequence. Indeed, it can be argued that in these cases the jury is still out on whether or not full recovery has been achieved. The chapter by Martin S. Alexander and Philip C. F. Bankwitz examines the attempts of French society and her armed forces to recover from the debilitating debacles of two colonial experiences in Indochina and in Algeria. Alexander and Bankwitz describe the monumental task confronting the civilian leadership of France in the years 1958–66: to reverse the image of a deeply humiliated military and, in addition, to initiate the type of structural reforms that would reestablish the officer corps' professionalism and obedience to civilian rule. In this case recovery has to be assessed, not by tabulating military victories, but by evaluating progress toward these objectives. Central to this undertaking was a realization of France's overextension and the need to reorient military strategy toward a set of goals consistent with her shrinking defense perimeter. De Gaulle's unique contribution lay in his comprehensive articulation of a new mission: a mission that reflected the transition from the politicized warfare of colonial conflicts to the more traditional nation-state conflict of the European theater. And although the French armed forces have yet to be "tested" in their new role, there are at least two reliable indications of French recovery from the colonial experience: the unquestionable professionalism of the officer corps and the acceptance, by all major political forces, of the crucial role of the force de frappe in France's strategic doctrine.[4]

The contemporary focus of the remaining three chapters raises important issues concerning what constitutes evidence of recovery from defeat where insufficient time has elapsed for the historian to be able confidently to establish the boundaries of the defeat-recovery cycle.

Amos Perlmutter examines the Israeli experience in light of the 1967 and 1973 Wars. Over the course of nearly twenty years (1947–67), the Israeli armed forces registered such an impressive record of success in the face of staggering odds that their failure to win a comparably decisive victory over Egypt in the years after 1967 has raised serious questions about Israel's military effectiveness. The myth of Israeli invulnerability and invincibility was shattered in 1973 by Egypt's early successes in the battlefield. Although the Israeli ability immediately to recover the military initiative and advantage was impressive, the nation's leaders were forced to face forthrightly for the first time the always intractable problem of how to convert military success into a stable and peaceful postwar regional order. The Israeli case shows clearly that battlefield prowess does not easily or necessarily transform itself into successful grand strategy and that too great a focus on winning a military victory can create conditions that may be inhospitable to the creation of long-term solutions.[5] The problem is compounded by the fact that, unlike the other cases examined in this collection, the Israeli "margin of error" is exceedingly small; even a minor setback has potentially catastrophic consequences.

Andrew F. Krepinevich, Jr., examines another case where the full nature and extent of recovery from defeat is still not apparent. The reasons why the United States armed forces failed to achieve victory in Vietnam remain a hotly debated topic to this day, even after the Persian Gulf War (January–March 1991) replaced Vietnam as "the war" for millions of Americans. Bureaucracies, especially military ones, are notoriously conservative institutions, and they can display an ability to ignore what they view as unpleasant or unrepresentative events. Krepinevich argues that the United States army, the focus of his chapter, still has to acknowledge that it experienced defeat in Vietnam and that it shares some of the responsibility for that outcome. The army seems to regard its part in the war in Southeast Asia as an anomaly that was appropriately not allowed to influence its role during the Cold War and should not influence the role it will play in whatever security commitments eventually emerge as part of the (yet to be defined) New World Order. In effect, the army has decided to recover psychologically from defeat by blaming the civilian leadership and by redefining the problem. It has concentrated on becoming very good at conventional war—although the brevity of the ground engagement during the Gulf War does not provide any conclusive evidence— and consigning whatever it might be able to learn about counterinsurgency warfare from its Vietnam experience to the scrap heap. As long as the army fights only conventional wars, such a perspective may not cause too many problems. But, as Krepinevich warns and as events in Somalia and other places (Bosnia, for example) might yet prove, counterinsurgency warfare skills may become the skills most in demand in the 1990s.

The final chapter by Gerald Segal, on China's recovery from her 1979 confrontation with Vietnam, stresses the importance of nonmilitary measures in achieving the desired objectives. In fact, Segal views the process of recovery as an essentially political process, echoing Peter Paret's discussion of the Prussian recovery after Jena. For many years, China was a "black box" to the rest of the world. Her virtual isolation for almost thirty years provided few—if any—reliable clues on her political, economic, and strategic objectives, let alone her decision-making processes. This lack of information puts a premium on examining those events that can open a window onto the thinking and behavior of the Chinese leadership. China's unsuccessful border war with Vietnam revealed major shortcomings in the armed forces. It was these very weaknesses that reformist forces took advantage of to shape a very different China. In the decade following the Vietnamese debacle, the country witnessed a series of sweeping military, domestic, and foreign policy reforms. While military reforms, particularly those pertaining to the state of professionalism in the armed forces and the acquisition of advanced technology, were in themselves impressive, they cannot be understood outside the context of domestic reforms. As Segal argues, China's wise decision to stress the civilian basis of recovery rather than the modernization of the military at the expense of the society at large contributed more than anything else to the durability of reform. Moreover, this civilian-based recovery had important effects on the country's international standing, enabling her— during the final phases of the Cold War—to achieve peacefully many of her foreign policy objectives. Although many issues remain unresolved, it can be argued that China is better prepared to confront the major challenges that lie ahead than at any time since the Communists came to power.

The purpose of this volume is twofold: first to suggest some new categories and concepts for thinking about familiar military events, and second to indicate that the cycle of defeat and recovery is an infinitely complex process that cannot easily be straitjacketed into a set of universally applicable criteria. The variety of the case studies presented here indicates some general directions; it does not provide a detailed road map. This word of caution is not an argument against the construction of a set of theoretical propositions; in fact such an undertaking is as challenging as it is desirable. What it argues against is a set of propositions based on a dubious commitment to historical relevance. If this collection alerts the readers to these concerns, the efforts of both the editors and the contributors to this undertaking will not have been in vain.

1

The Royal Navy's Recovery
after the Early Phase of the
American Revolutionary War

R. J. B. Knight

A sharp dip in the seemingly relentless rise of British naval power took place in the first years of the American Revolutionary War; indeed, during 1778 and 1779 the Royal Navy was almost completely ineffectual. Within eighteen months of the defeat of the army at Saratoga, it fought an inconclusive action off Ushant that many considered a defeat and experienced major political court-martials of the two admirals involved in the battle, Augustus Keppel and Hugh Palliser, which split the service and the country. The Spanish entered the war in early 1779, and in that summer a combined Franco-Spanish fleet entered the Channel intent on invasion. Lord Sandwich, the first lord of the Admiralty, was under severe political attack in both houses of Parliament, while Lord North, the prime minister, had lost the will to make decisions and wanted nothing more than to resign; only the king's obduracy prevented the government from collapsing. The sixty-six ships of the Franco-Spanish fleet were faced by the forty-two ships of the Western Squadron (including old ships commissioned "for summer service only"), commanded by a benign but lackluster sixty-three-year-old admiral, Charles Hardy. Worried about the threat to Ireland, he took his fleet well to the west. The French-Spanish fleet came between the British fleet and the English coast, panic ensued at Plymouth, and the citizens fled. One member of the government, Lord George Germain, commented: "I think we have more reason to trust in Providence than in our Admirals."[1]

The reasons for the loss of the thirteen American colonies and the

I am grateful to Daniel Baugh, Michael Duffy, Brian Lavery, and Paul Webb for their comments on this chapter.

part that naval failure played in it have been attributed to various causes over the years. The late Victorians were either dismissive or bewildered. Sir John Knox Laughton and Admiral W. M. James, for instance, influenced by the victories of the Seven Years' War and subsequent British success against Revolutionary and Napoleonic France, blamed the corrupt administration of Lord Sandwich or made general observations on naval malaise. One book, published as late as 1987, dismissed the problem by stating, without analysis, that Sandwich's administration was "abysmally inept."[2] R. G. Albion's imaginative but erroneous theory on the effect of the lack of American masts on the British navy, published in 1926, has made its mark.[3]

Since the 1960s, the naval war has been the object of much more systematic attention, beginning with Piers Mackesy's book *War for America* (1964), which placed the loss of the colonies in a domestic and European context. Jonathan Dull's book on the French role in the war is immensely valuable in assessing French intentions and resources, while Hamish Scott puts the conflict into the context of foreign policy.[4] David Syrett has analyzed the naval war in American (though not Canadian) waters; recent studies by Nicholas Tracy and Michael Duffy of the periods before and after add pieces to the puzzle.[5] Nevertheless, a comprehensive work on the strategy and resources of the European and naval war has yet to be written, and in particular on that vital ingredient of seapower, naval manning.[6] The only piece of work to fill that vacuum is a short article by Daniel Baugh, in which he challenges the idea of the often noted disadvantage created by England's lack of allies in Europe to distract France.[7] Those allies cost money in subsidies, and that money was better spent on the navy. He noted how strong the British navy was by the end of the war, and how the French and Spanish navies had declined. If this is a correct view, and he presents a strong case, one may certainly ask why, with so many inherent strengths, did the British navy perform so poorly at the start of the American War?

This chapter does not focus on strategic analysis, for England's recovery in this war owed nothing to a revamped strategic approach. Suffice it to say that Sandwich's naval direction reflected his as well as the government's weak position. The most frequent criticism leveled at him is that he was overly protective of home waters, and that public opinion, shaken by the invasion threat after 1779, influenced his decisions. This fear kept fleets at sea well into the winter months, punishing the ships. It also delayed the dispatch of vital fleets to North America in 1778 and

1779 and to the West Indies in 1780, and it weakened the defense of the Mediterranean garrisons. While the war was never going to be won in the Channel, Sandwich was too conscious that it could have been lost there.[8] Strategic boldness comes only from confidence and mutual trust among members of the government. Yet internal tensions, particularly between Sandwich and Germain, made it impossible to risk the occasional reverse. Underlying this political weakness was the fact that Britain faced a new political phenomenon, large-scale colonial rebellion.

Opposition in Parliament and country led Lord North and his colleagues into a series of unprofitable political calculations and gambles between 1774 and 1777. Public opposition led to a hesitation to build ships, large and small, to strategic indecision, and to damaging tension between the government and the professional navy. After French and Spanish intervention in 1778 and 1779, and once the conflict was translated into the more traditional pattern of a European war, it became clear that Britain's navy was very badly outnumbered. This realization jolted the government out of its slumber. The government made a series of radical decisions concerning shipbuilding and the application of new technology. It committed resources in hitherto unknown amounts; its naval administration became reinvigorated. This dramatic initiative gave the Royal Navy such momentum that, not only did it finish the American War strongly, but it won a lead over its rivals which was to last, threatened though it was by Napoleon, throughout the age of the wooden ship.

It is best to approach this problem by seeing two mobilizations in this war. The first one, in 1775, launched the war in North America. It was too little. The second, to meet the French and then the Spanish threat, began at the end of 1777. This was too late.[9] North's government totally underestimated the naval requirements in North America in the first years of the war and tried to get away with a quick military victory and minimal naval expense. One of the first acts of any British government at the first sign of trouble would be to order frigates and sloops to be built on the London River, but North failed to do this.[10] Due to the politically sensitive nature of this rebellion, the government lacked the political will to act. The rebellion was not quashed, and all available frigates had to be sent to North America, leaving none in home waters. Yet in spite of intelligence reports that the French were investing in preparing their navy, North still insisted on budgetary stringency.

There were a number of reasons why Sandwich lost the early argu-

ment for increased navy funding and why North insisted on budget stringency. In September 1772 North wrote to Sandwich, "This is the time, if
ever there was a time, for a reasonable and judicious economy." He then
added prophetically: "It must be owned that we suffered a little from the
unprepared state in which we were at the opening of the last two wars;
but then, our resources, our credit, and the length of our purse, which
had been carefully managed during the preceding times of peace, carried
us through with glory and success." Sandwich could only reply, "I do not
entirely see it in the same light." [11] The naval debt had risen because of the
Falklands, and there was a good deal of relaxation because of the diplomatic success of that crisis. Intelligence reports depicted a weak French
navy and discounted the Spanish one. The government became distracted
by Spanish-Portuguese tension. Besides, Sandwich was spending a good
deal of money building up the reserves of naval stores. Nevertheless,
North hesitated and held onto this position after the news of Bunker Hill,
when intelligence reports were quite clear that the French were rearming.[12] Through these years Sandwich was forced to make savings. The
dockyard workforce was cut in March 1774. Sandwich had further to
force the pace of change by the attempt to implement task work (payment
by tasks rather than by the day). This, in turn, led to a hard-fought strike
in all the dockyards except Deptford. By July 1775, of course, the situation had changed dramatically and the Navy Board ordered the yards to
hire any shipwrights they could get. Despite this policy, the yards could
not obtain the shipwrights they need for the rest of the war.[13] Orders for
ships of the line came to a virtual halt until late in 1777. With the average
three years' building time, it meant that ships of the line were being built
at a rate of only two a year for the first three years of the war. It was just
not enough.

When the news of the defeat at Saratoga reached London, North
realized how unprepared he was. He had gambled on a quick victory.
He tried to resign, but the king would not agree. Then the pendulum
swung the other way. Political weakness turned almost to political panic.
Lord Sandwich was now pressed to build, and his administration, with
the newly appointed and energetic Charles Middleton as controller of the
navy, approached the shipbuilders. A greater number of shipbuilders than
ever before were engaged to build naval ships. Eighteen ships of the line
were ordered in 1778–79; but it took a minimum of three years to build
such a ship. True, five already standing in frame in the royal yards were
finished off and launched within eighteen months, and merchant yards

quickly produced fifty-five frigates and sloops (25,750 tons) in those two years. The initial hesitation was rectified but the delay caused a critical shortage of ships in America during the only period in the war when the rebellion could have been stifled by blockade or crushed by greater naval and military force.

A French naval challenge was yet another obstacle that Lord North's government did not take seriously enough, early enough. Apart from a brief period in the previous century, France had presented no real naval threat; but after 1763 the French navy was given a chance to mobilize resources on a larger scale. Although the duc de Choiseul failed to meet the targets he set himself, he did succeed in some important infrastructural reforms. In addition, the Toulon and Brest dockyards were improved, as was the administrative structure of the French navy, and he increased the fleet to sixty-seven battleships, although he had hoped to reach eighty. Even so, by 1770, when Choiseul fell from power, France did not feel able to challenge British strength. This explains why in the Falkland Islands crisis of 1770, the Bourbon powers held back. Nevertheless, during the 1760s warship construction in Britain and France proceeded at the same rate and in the 1770s the relative strengths of the two nations were converging.[14]

After the war, lessons were absorbed. In a revealing analysis, a clerk from the Navy Board, Charles Derrick, wrote in his *Memoirs of the Progress of the Royal Navy* that Britain, in order to meet the challenge of both France and Spain, needed to maintain a hundred ships of the line in good condition. In order to do this, it was calculated during the 1780s that Britain had to build or make a large repair on ten ships of the line every year.[15] Table 1 shows that the building of ships of the line did not approach this level until 1782; nor, during these years, were enough ships of the line given major repairs, as table 2 reveals. Taken together, these two tables make it clear that, prior to the French intervention in March 1778, Sandwich was losing the battle against decay; and after 1775, when the Bourbons started rearming, he was losing the battle before it was fought. The navy averaged a gain of three new ships of the line a year between 1771 and 1778.[16] During the same period it broke up ships at a rate of 3.5 a year. It was only making 2.5 major repairs a year. If Sandwich inherited 86 ships in 1771, as he said, then matters were serious indeed, as he realized a year later. In short, the expansion of the fleet from the time of the Seven Years' War, both in numbers and in the size of ships, created a navy that the Admiralty could never maintain properly.[17]

Table 1.   British Naval Ships Launched by Year

	Ships of Line		50 Guns and Under		Total	
Year	No.	Tonnage	No.	Tonnage	No.	Tonnage
1771	1	1,650	3	1,222	4	2,872
1772	3	4,699	1	35	4	4,734
1773	2	3,579	7	4,570	9	8,149
1774	7	10,357	9	7,261	16	17,618
1775	4	6,220	4	2,374	8	8,594
1776	2	3,028	18	6,103	20	9,131
1777	2	3,309	17	7,140	19	10,449
Subtotal	21	32,842	59	28,705	80	61,547
1778	2	3,259	24	10,987	26	14,246
1779	3	4,900	29	14,863	32	19,763
1780	5	7,167	19	12,132	24	19,299
1781	7	10,192	21	13,967	28	24,159
1782	10	15,194	18	11,509	28	26,703
1783	6	9,793	20	12,839	26	22,632
Subtotal	33	50,505	131	76,297	164	126,802
Total	54	83,347	190	105,002	244	188,349

Source: Basic information is taken from *Admiralty Abstract of Progress Books, 1759–1821* (Public Records Office, ADM 180/6–9) and Navy Board warrants to the dockyards, 1774–1781 (Public Records Office, ADM 95/95 and 96). Supplemental information comes from the list of ships in Brian Lavery, *The Ship of the Line,* 1: 178–86, and J. J. Colledge, *Ships of the Royal Navy: An Historical Index* (Newton Abbot, 1969), vol. 1.

In addition to the problem of an inadequate number of ships, the lack of political will ensured that Sandwich could not "pick a . . . fighting admiral."[18] There can be no doubt that the professional officer corps was divided into factions for a good deal of this war and that this factionalism was extremely serious, was at the center of the political stage at a crucial time, and excluded professionally respected admirals from vital commands. This divisiveness caused a morale problem in the Channel fleet for four years and for much of the time, though perhaps for slightly different reasons, in the West Indies fleets. At the time, Sandwich was seen as the cause of this divisiveness; was it his fault? If so, why?

It has been suggested that the prime reason why distrust was endemic

Table 2.   British Naval Ships of the Line

	Major Repairs			Broken up/Sold		
Year	100+90 guns	74 guns	64 guns	100+90 guns	74+70 guns	66/64/60 guns
1771	—	—	—	—	—	3
1772	—	1	2	—	1	6
1773	—	2	1	—	1	3
1774	—	—	2	1	2	4
1775	—	3	—	1	3	1
1776	—	1	—	—	—	1
1777	—	1	2	—	—	1
1778	1	3	—	—	—	—
1779	1	—	—	—	—	—
1780	1	3	2	—	—	—
1781	—	2	1	—	—	—
1782	2	7	4	—	—	—
1783	—	3	—	—	—	—

Source: Basic information is taken from *Admiralty Abstract of Progress Books, 1759–1821* (Public Records Office, ADM 180/6–9) and Navy Board warrants to the dockyards, 1774–1781 (Public Records Office, ADM 95/95 and 96). Supplemental information comes from the list of ships in Brian Lavery, *The Ship of the Line*, 1: 178–86, and J. J. Colledge, *Ships of the Royal Navy: An Historical Index* (Newton Abbot, 1969), vol. 1.

Note: Judgment is needed on what constitutes a major repair, for there is considerable discrepancy between the dockyard officers' classification of a "large," "middling," or "small" repair and the cost of each repair which is so assiduously listed in the Abstract of *Progress Books.* For the purpose of this analysis, repairs costing over £20,000 for a first rate, £15,000 for a 74-gun ship and £10,000 for a 64-gun ship have been taken as a major reinvestment, representing approximately one-third of their building cost. This method gives the Sandwich administration the benefit of the doubt; I doubt whether these criteria would have satisfied Charles Derrick and his mentor, Charles Middleton, for the sums spent on the repair of each ship in the late 1780s were far in excess of these figures. Those ships that were guardships during the peace did not need major repairs since a steady amount of money was spent on them annually and their maintenance was high on their captains' list of priorities. The table for ships broken up or sold is taken directly from a Navy Office Paper, dated 27 December 1781, of ships broken up between 1771 and 1781 (*Sandwich Papers*, 4: 306).

in this period was that Sandwich was unique in the eighteenth century—that is, a civilian politician who knew as much about the navy as the admirals themselves. But the picture of the admirals against one man, albeit a civilian, is inaccurate. For if the admirals did not like Sandwich, they liked each other even less. There was, for instance, intense personal rivalry between the leading admirals of the day: Richard Howe never got on with Keppel; no-one trusted George Rodney; Rodney was suspicious of everyone; Commodore George Johnstone hated Howe; Howe had not talked to Lord George Germain since 1758.[19] The politics of the period, intense in the country as a whole, aggravated the situation, but rarely initiated problems. Personal and professional rivalries were at the heart of this widespread distrust.

These rivalries focused on three main areas of conflict. The first two related problems were the squabbles over prize money and the equivocal position of flag officers appointed to foreign stations. These problems were not unique to this war, indeed, they were a constant irritant throughout the eighteenth century. Rodney had longstanding differences with Marriot Arbuthnot and with Peter Parker on both counts, while he fell out with all and sundry over the St. Eustatius booty. The protracted quarrels, in particular in the months before the Battle of the Chesapeake, were particularly damaging.[20] The friction that was generated by imprecise orders and boundaries, particularly on the North American station, which was commanded largely by second-rate men, was at times laughable. The third area of conflict involved promotion and appointment, decisions that were controlled by the first lord of the Admiralty. By the time of the war with America, the options available for rewarding those at the top had become too few and were too complex. If one admiral was rewarded with the sinecure of a lieutenant-generalship of marines, or ordnance, or the treasurership of the navy, the others were alienated. The traditional view of the problems with promotion, proffered by Sir John Laughton, is that Sandwich was corrupt.

The patronage system has been defended in detail in the last few years. The most recent and persuasive defense argues that from a professional standpoint the checks and balances in a system of personal recommendation worked extremely well, particularly when naval officers knew everyone's professional reputation very well, when captains were on the lookout for talented junior officers, and when a minimum standard of competence had been established by lieutenants' exams after six years at sea. In the mid-eighteenth century, when the navy was administered by a naval officer of unassailable professional reputation, a sound fortune, and a strong general political base, this was undoubtedly the case. Sustained analysis by N. A. M. Rodger demonstrates patronage in a similar fashion to that most respected of sea officers and first lords, George Anson, that he resisted political influence as much as he could and that in applying strict standards Sandwich "was deviating sharply from what was expected of him by colleagues and contemporaries."[21]

The other part of the problem was that naval officers were too near the heart of politics. Let us take the 1780 Parliament where sixteen naval officers had seats, later increasing to eighteen. Six were flag officers (Rodney, George Darby, High Pigot, Keppel, Howe, Molyneux Shuldham). The others included Governor Johnstone, Lord Mulgrave, Lord Robert Manners, George Berkeley, Edmund Affleck, and George Elphinstone. However, only three naval officers can be described as active politicians closely associated with a tight-knit political group during the course of the war.[22] Mulgrave and Palliser (after the court-martial of 1779, of course, a political embarrassment) were closely associated with the administration, and Keppel with the Rockingham group. Naval officers sought to become members of Parliament for a variety of reasons, not only to cultivate political connections, but also for local and family reasons. A seat in Parliament could be useful, for instance, because an admiral on a foreign station might be vulnerable, particularly to mercantile interests in disputes over convoying, prizes, or pressing for crews. Opposition to Sandwich therefore was very far from being "party" based. When things were going badly, Parliament provided the venue in which professional and political grievances could directly challenge the authority of the first lord of the Admiralty. If he was not strong politically, then this weapon could be very effective; and Sandwich was far from strong as a political figure.

There was a contradiction in Sandwich's personal political position within the North government. First, he controlled a number of seats in Huntingdon and elsewhere that were of importance to the prime minister. Yet Sandwich rarely called the tune; his need for money, and thus the need to cling to office, led to few risks. Lord Sandwich's position was very weak; he was neither political nonentity, nor a political liability; but he was, apparently, politically irreplaceable. The underlying problem was the weakness of Lord North's leadership, which was nonexistent at times.[23] At two or three crucial points, only the will of the king kept the administration in office. Antagonisms among the members of the administration reflected North's lack of will. The cautious Sandwich was too

often opposed by Germain, who was always pressing for attack. Throughout, it seemed that Sandwich never felt himself strong enough to deal
"straight" with his admirals. A number of scholars have picked up instances or periods when he was particularly weak. Professor Baugh notes
his lack of clear direction in the politically sensitive Howe command in
1776. Piers Mackesy picks up his tendency to hide behind his cabinet
colleagues or to take the easy path of conversations, meetings, and private letters rather than issue clear written orders in the hectic summer
of 1779.[24]

Who would Sandwich have had to deal with? First, there were the
capable, though conceited, admirals—Howe, Keppel, and Rodney. At
various times they were all on good terms with Sandwich. All of them fell
out with him eventually, but then all were difficult men. Add Augustus
Hervey, the Earl of Bristol, Samuel Barrington, Johnstone, and Pigot to
the list of able men who opposed him. Second, there were the pensioners
or makeweights—the compromise, or just bad, appointments. These included Hardy, Thomas Graves, Darby, Arbuthnot, Shuldham, and, worst
of all, James Gambier. Third, there were the able administrators who were
political lightweights: Middleton, Palliser, and Maurice Suckling came
into this category. Only Samuel Hood and Richard Kempenfelt were able
and kept their nonaligned reputation. It was this last group upon which
Sandwich relied. With the weakness both of his own position, and of the
government's, the result was weak or cautious strategic decisions and
nonthreatening appointments.

Yet the British did recover from this position of weakness. To date the
turnaround in 1781 is perhaps to fly in the face of obvious facts. With the
French and Spanish cooperating effectively, this was a successful year for
them: Pensacola and Tobago were captured; Minorca fell; and the strategic advantage sprung by de Grasse at the Chesapeake led to the capture
of Cornwallis's army at Yorktown. This vital defeat broke the British will
to see military victory in North America as a possibility. The French then
set out to press home their advantage with attacks on Jamaica and India.

The war became a real test of stamina. Who had the reserves of ships,
of men, and of money? The French concentrated larger and larger numbers of workers in their dockyards by means of impressment and by the
use of convict labor, and they did not use private builders as a matter
of course. By contrast, the six British royal dockyards built fewer and
fewer warships as the century progressed, and once war began, they built

Table 3.   British Naval Ships Launched by Yard

	Royal Dockyards				Merchant Yards			
	Ships of Line		50 Guns and Under		Ships of Line		50 Guns and Under	
Year	No.	Tonnage	No.	Tonnage	No.	Tonnage	No.	Tonnage
1771	1	1,650	1	302	—	—	2	920
1772	3	4,699	1	35	—	—	—	—
1773	2	3,579	—	—	—	—	7	4,560
1774	3	4,389	3	1,896	4	5,968	6	5,365
1775	3	4,606	4	2,374	1	1,614	—	—
1776	2	3,028	7	2,622	—	—	11	3,481
1777	2	3,309	3	1,093	—	—	14	6,047
Subtotal	16	25,260	19	8,322	5	7,582	40	20,373
1778	2	3,259	4	1,413	—	—	20	9,574
1779	3	4,900	4	1,591	—	—	25	13,272
1780	1	1,370	4	3,533	4	5,797	15	8,599
1781	4	5,995	5	2,830	3	4,197	16	11,137
1782	4	6,065	—	—	6	9,129	18	11,509
1783	—	—	1	1,050	6	9,793	19	11,789
Subtotal	14	21,589	18	10,417	19	28,916	113	65,890
Total	30	46,849	37	18,739	24	36,498	153	86,263

Source: Basic information is taken from *Admiralty Abstract of Progress Books, 1759–1821* (Public Records Office, ADM 180/6–9) and Navy Board warrants to the dockyards, 1774–1781 (Public Records Office, ADM 95/95 and 96). Supplemental information comes from the list of ships in Brian Lavery, *The Ship of the Line,* 1: 178–86, and J. J. Colledge, *Ships of the Royal Navy: An Historical Index* (Newton Abbot, 1969), vol. 1.

very few indeed, concentrating instead on repairing and refitting existing ships. The merchant yards in the south of England took almost all the strain of building (see table 3).[25]

The rate of building increased as the war continued. Between 1778 and 1783—the years of the European war—an average of twenty-seven new ships were added every year to the British navy, a total of 126,802 tons during these years. The peak was reached in 1781 and 1782, with 24,159 and 26,703 tons, respectively. It is worth noting the construction figures of smaller ships, those of 50 guns and under, because the British were desperately short of frigates and sloops in the years before the European powers entered the war. Between 1778 and 1783, 131 were built, just under 76,000 tons. Of these, only eighteen were built in the royal dockyards—13 percent by numbers of ships and tonnage; the rest were built by the private sector. At the same time, the British line-of-battle fleet reached maximum strength in September 1782, and French intelligence was well aware of it. And perhaps more significantly, on the larger slips of the Thames, Medway, the Essex rivers and the Solent, a further thirty ships of the line were under construction.

It was this capacity for regeneration which Charles Vergennes, the French foreign minister, recognized in mid-October 1782, when he wrote to the French ambassador in Madrid, to persuade the Spaniards to make peace:

> The English have to some degree regenerated their Navy while ours has been used up. Constructions have not been at all equivalent to consumptions; the body of good sailors is exhausted and the officers show a lassitude in war which contrasts in a disadvantageous manner with the energy which not only the sailors but the entire English nation eagerly manifests. Join to that the diminution of our financial means which are limited by reason of the usage which has been made of them. That inconvenience is common, no doubt, also to England, but her constitution gives her in that regard advantages which our monarchical forms do not accord us. She will pay dearly for money, but she will find more easily than us all that she needs.[26]

British capacity was further helped by two technological innovations, arguably the only effective ones in the century. Both were the result of decisions made in 1778 when the government had its back to the wall. Charles Middleton enthusiastically argued for these innovations, which were both successful gambles that were effective from 1780 onwards.

The first was the adoption of copper sheathing of ships' hulls. The search for an effective sheath, first, to inhibit the growth of barnacles and seaweed which slowed the ships and, second, to stop the depredations of the *teredo navalis,* had been continuous for more than a hundred years. Softwood or elm sheathing, covered with a sulfurous composition, was used prior to this time, but it had the disadvantage of having to be renewed every three years. Since the late 1760s, experiments with copper had been carried out on small naval ships to find an effective means of stopping the galvanic corrosion of the iron fittings in ships caused by the contact of copper with sea water. By the beginning of the European war, tarred paper between copper and the heads of the iron bolts had had some good results. In 1778, at a meeting between the king, Sandwich, and Middleton, the decision was made to copper the whole fleet. The majority of the fleet was coppered in two years, which was a considerable achievement. In 1780 alone, 42 ships of the line were given the new sheathing. By early 1782, by the end of Sandwich's administration, 313 ships (82 capital ships, 14 of 50 guns, 115 frigates, and 102 sloops and cutters) had been coppered. Though the French followed suit, they were delayed by technical problems with copper nails, some difficulties in distributing materials, and later shortages of copper. Only half the French ships that went to the West Indies in 1781 were coppered, while only one Spanish ship was coppered in this war.[27]

Advantages came in different ways. Ships no longer had to be docked when refitted, a process that averaged between four and five months for a seventy-four-gun ship; now they could be turned round within weeks. Just as important, once the coppering had taken place, it enabled the dockyards to clear their docks for repairing. Table 2 shows that thirteen ships-of-the-line were given major repairs in 1782; at the same time as the building capacity crescendoed in 1782, the repairing program hit its peak.

At sea, the impact of copper sheathing is more difficult to assess, but the admirals were enthusiastic. Kempenfelt was quoted as saying that "25 coppered ships of the line were enough 'to tease' the combined French/Spanish fleet in the Channel in 1779." Rodney attributed much of his success in capturing six Spanish ships-of-the-line in the Moonlight Battle in 1780 to coppering. Confidence grew in the latter part of the war, and, as one historian notes when writing on copper sheathing, "in war . . . believing in one's own superiority is not to be taken lightly."[28] Another study demonstrates that in the final years of the campaigns in the West Indies the battle was as much to keep the ships afloat and at sea as fighting the French, and the superiority of English coppered ships of the line was a key factor in success in these last years. Coppering came to be taken into strategic as well as tactical thinking.[29]

Yet the decision was a gamble, because of the technical difficulties that were known but not fully understood despite the experiments in the 1760s and 1770s. The galvanic effect of the sea water and the copper on the ships' main iron bolts slowly continued during the war, in spite of the improved methods used to prevent it. Although the copper provided a clear advantage during the fighting, this galvanic action wore away at the iron bolts and in one incident, four ships of the line were lost.

In September 1782, when the French ships captured at the Saints, the *Ville de Paris* (110) and *Glorieux* (74), and the English *Ramillies* (74) and *Centaur* (74) foundered in a severe Atlantic storm while returning from the West Indies; 3,500 British seamen lost their lives, more than in ten years of naval warfare. Moreover, it was a financial gamble. Although the copper sheathing immediately added between 10 and 15 percent to the cost of a ship, the correspondence in 1778–79 does not mention the financial implications of this decision. In the 1780s, there was a further immense cost: Thomas Williams, the principal copper contractor, developed a compound metal bolt that contained copper and thus stopped the corrosion. Although Charles Middleton at the Navy Board was skeptical at first, it was decided in August 1786 that all ships had to be given the new bolts.

The other technological advantage that the British had was the carronade. This was a powerful, light, wide-caliber gun, very effective at short range. The ball fitted tightly into the barrel and used less power, and thus the gun could be built more lightly (a 32-pound carronade weighed the same as a 6-pound long gun). They were adopted in 1779, despite resistance from naval officers. After several successes, all such resistance was overcome in 1782 when a 44-gun frigate (the *Rainbow*) encountered a French 38-gun privateer (the *Hebe*) that was more than 200 tons larger, and caused her to surrender after a single broadside at short range. Carronades were rarely solely used, for an enemy could keep its distance, but they changed the pattern of naval warfare, and close action became more common. Eventually, of course, the French copied the gun, but they did not use it effectively before the end of the war.[30]

Some of this effort was translated into British success at sea by 1782. Rodney brought discipline into the West Indies fleet and his thirty-six ships beat de Grasse's thirty-three at the battle of the Saints in April of that year. Hughes kept Suffren at bay in India. It was clear, too, that the blockade of Gibraltar, the key objective upon which the Franco-Spanish alliance was founded, was so loose that Darby, in relieving the siege for the second time in March 1781, was scarcely molested; the Spaniards' final massive effort failed and Howe carried the third relief in October 1782. The British thus had a few late cards to play in the protracted peace negotiations, which culminated in an agreement in March 1783.

Thus Britain's financial and naval resources came near to neutralizing the political and strategic hesitancy of the period before 1778. In contrast to the financial squeeze of the prewar years, an unprecedented level of funds was thrown at a political problem—the challenge to naval superiority. Between 1778 and 1781, France regained the prestige and some of the territory she had lost in the Seven Years' War and Britain lost the American colonies. Britain could have lost more of her empire to the French, for Jamaica and India were at risk, but by the end of 1781 superior financial and shipbuilding resources put the initiative back into British hands. Lord North's budgetary stringencies of the early 1770s and the crucial alliance with Spain gave France a three-year "window of opportunity" in 1778; and it is not entirely coincidental that three years was the contract period for a private shipbuilder to build a seventy-four-gun ship.

What comes into play here is also the nature of government. France, authoritarian in structure and style, had in this war fewer problems with policy and choosing objectives, yet her resources were less organized and controllable. During this war, Britain's policy- and decision-making processes were ineffective, though her policy was supported by increasingly well-administered and potentially stronger resources. The North government wove its way through these years, as if through a maze without a map, battered by opponents of the idea of the war itself and by a navy that was weakened by internal struggles.

Matters thus came down to money and to credit. Professor Baugh advances the figures of 81 percent borrowed, and only 19 percent covered by tax revenues.[31] Britain carried a huge naval debt three years into the peace, which Parliament funded with scarcely a murmur. The ability of eighteenth-century Britain, or in John Brewer's phrase the "fiscal-military state," to shoulder large and increasing debts underpinned this great effort.[32] There can be no doubt of the British naval and economic advantages. British naval strength, in an elegant phrase of William McNeill's, was "supple and effective." It was supple because of the unrivalled credit system based on the Bank of England which fuelled the private sector, eager to make profits by producing a British fleet. It was effective because of what McNeill less elegantly described as the "feedback loop," for the naval industry was particularly rooted in the economy. "Naval power and expenditure reinforced commercial expansion while commercial expansion simultaneously made naval expenditures easier to bear."[33]

The French attempt to convert resources and money, almost overnight, from the army to the navy could not be long-lasting. In Professor Baugh's words: "No amount of French or Spanish money could quickly augment the pools of trained seamen and shipwrights, or suddenly create well-equipped dockyards and overseas bases and well-tested administrative procedures, or secure high-quality naval stores."[34] Or in the words of a pamphleteer of 1782:

> Naval strength is not the growth of a day, nor is it possible to retain it, when once acquired, without the utmost difficulty, and the most unwearied attention. The English have proved by their conduct, for almost two centuries, the firmness and steadiness of their naval character. Whereas the maritime enthusiasm of the French has only occasionally taken place, and does not seem consistent with the natural bent and genius of the people.[35]

2

. . . .

The Recovery of Prussia
after Jena

Peter Paret

Wars, whether they are won or lost, always have their greatest impact, not on the armies involved, but on the societies for which these armies fight. In turn, the impact of war on society will, in the final analysis, determine the character and mission of its armed forces from then on. If a service wishes to respond to its experiences in a conflict just past and to the new political realities that the war has created, it can do so only in interaction with its nonmilitary environment. Reorganization, new weapons, rearmament—these come about as the result of negotiating with political and economic interests, of accommodation with the conditions of society—or as the result of seeking to change society. Even within the service, significant changes in such matters as the design of weapons or the development of doctrine usually have a political dimension from the start. And once carried out, such a change—no matter that it may occur in a narrowly technical area—will always have political consequences.

Although every author writing in this volume presumably has something to say about the links between armed forces, society, and politics that make possible the process of recovery from war, only a few, I should think, need to bring the nonmilitary elements of their particular episode as deeply into their analysis as I shall have to do. As the chapter title indicates, I am concerned with the recovery, not of the Prussian army, but of Prussia herself. In fact, if we want to be very precise, we may find that in one respect the title remains slightly ambiguous. From the perspective of military and political power we can certainly speak of a Prussian recovery after Napoleon overran the country in 1806. But what happened to the army and the state in those few years between Prussia's collapse and the final defeat of Napoleon goes far beyond recovery to very considerable changes in many of the basic elements, not only of the army, but of the state and society as well. And significant change in any one of these areas could not have occurred without change in the rest. The reform of the Prussian army and the reform of society, of its political structure and its economic practices, were parts of the same process. We may think that the specific forms this interaction took turned out in the long run to be a misfortune, even a tragedy, for Prussia and Germany. But it is also apparent that the ways in which civilians and soldiers cooperated—as the result of which Prussia was radically modernized while retaining an autocratic government—was the condition that at the time made the military reforms, the recovery of the army, so effective.[1]

Why did Prussia lose the War of 1806? The question needs to be asked because the reasons for her defeat, and what people at the time believed these reasons to be, determined the direction and nature of Prussia's recovery. After sketching out the main features of the war, I shall turn to the reforms themselves. What did they consist of? What were the central conflicts between those favoring innovation and their opponents who to greater or lesser extent fought to preserve the status quo? What was and what was not achieved? Finally, it may be useful to consider how these changes worked in practice in the wars of liberation, and what effect they had in the longer term.

I

After the two main Prussian armies were defeated on the same day—14 October 1806—in two battles fought a dozen miles apart, and after the subsequent surrender of troops and fortresses, even while other forces continued to resist, a view quickly spread that the war had been a political and social test, which Prussia had failed. Many contemporaries at the time and many historians since have seen the war as a confrontation between the traditional and the modern, in which the hidebound was measured against the enlightened and found wanting. On one side, France, modernized by the Revolution and Napoleon's military dictatorship; on the other side, Prussia, no longer the absolute monarchy of Frederick the Great, but still an autocratic state based on a hierarchic society, its essence most fully expressed in its outsized, expensive, highly drilled army. When they clashed, the old was vanquished by the new.

This interpretation, though persuasive, may be questioned on two counts. One refers to the discrepancy in the strengths of the two states. In 1806 France had nearly thirty million inhabitants, Prussia and her Saxon ally less than half that number. Population size is not an absolute indicator of power, but in this case it fairly accurately represents the differences in economic and human resources, administrative efficiency, and military potential. French superiority was such that even if Prussia had attained a comparable level of modernity by 1806, the likelihood of a French victory would still have been very great.

A second argument against the hypothesis that the defeat was caused by Prussian backwardness concerns the unusual strategic advantages the French enjoyed. After his triumph over Russia and Austria the previous year, Napoleon had not withdrawn his army to France but had left well over 100,000 men in southern Germany, deployed in a broad arc from the Rhine through Bavaria to upper Austria. Even before the conflict began, their forward positions were within 200 miles of Berlin, and an advance due north from Bavaria would directly threaten the Prussian field army's lines of communication to the capital, to East Prussia, and to Russia. Napoleon was able to retain his forces in Germany and to add to them quickly when war became imminent, because he had isolated Prussia diplomatically. Prussia had not joined the Third Coalition the previous year. Now neither Austria nor Great Britain was able to help her, and an alliance with Russia was only in the process of being concluded. Nothing, therefore, impeded the concentration of French power against a single target. Other differences between the two opponents unrelated to their political and social systems could also be mentioned, the most significant being that the Prussian army had not seen action since the 1790s, whereas the French was now the most seasoned and experienced army in the world.

To determine the validity of these objections to the thesis that Prussian backwardness caused her defeat, we must take a look at the war itself.[2] On 8 October, the French army, some 160,000 strong, moving in three main columns on a 35-mile-wide front, crossed the Saxon frontier and headed north. Opposing them were perhaps 120,000 Prussians, the greater part assembled between Weimar and Jena. On the 10th the Prussian advance guard was defeated and the French advance continued past Jena toward Naumburg, 150 miles southwest of Berlin. By the 12th Napoleon realized that the Prussians, instead of withdrawing in order to stay between him and Berlin, had left most of their forces west of Jena and that with every additional step he was outflanking them further. He swung a part of his army west toward Jena, which other units were approaching from the south, sent a second column north to Naumburg to envelop the Prussian flank and rear, and ordered a third column to occupy a central position between Jena and Naumburg in order to be able to intervene at either place if needed. Although his instructions were not completely carried out, he succeeded in rapidly concentrating an enormous amount of military power in an area from which he threatened the very existence of the Prussian state.

The generals opposing him, on the other hand, were neither motivated by a dominant strategic vision nor possessed of his energy and ruthlessness. On the morning of the 14th Napoleon attacked the Prussian divisions stationed west of Jena. He did not realize it, but his troops outnumbered their opponents by nearly two to one because during the previous day a large part of the Prussian army had at last marched north to escape the threatened envelopment. As the battle of Jena reached its climax, this second force tried to break through to the east and north, but was stopped after hard fighting by the much weaker French blocking force at Auerstedt near Naumburg. Although French tactics were much more flexible, the Prussians fought well on the whole, and French casualties were high—more than 25 percent killed and wounded at Auerstedt, less than that but still significant at Jena. The old system was by no means totally decrepit. What made the difference was the Prussian commanders' lack of personal initiative; poor coordination between their forces and, within each command, between infantry and cavalry; and weak, almost nonexistent, central control. Losing the two battles was serious enough, but what followed was far worse. The retreat was poorly planned and directed, soon supply and transport broke down completely, and many units dissolved into undisciplined masses intent only on escaping from the pursuing French. Fortresses and depots where troops might have been reorganized and reequipped were surrendered by their commanders, many of whom were aged officers who had been given these posts in lieu of a pension. On 24 October French troops entered Berlin; by the end of the month, they had reached the Baltic coast. Only East Prussia and Silesia remained for the time being under Prussian control.

To return to our question, was the war a test case? Some of the points I have just mentioned were used in 1807 and later by opponents of the reform program to argue that it was not. Mistakes had been made and the army's organization, equipment, and training were not equal to the task; but the true causes of the disaster were inferiority in numbers and resources, and the fact that Napoleon did not have to fight his way into Prussia from northern France and the Rhine, but could at once launch a decisive offensive from southern Germany.

The proponents of reform saw things very differently. They acknowledged that the French had enjoyed great advantages, but they denied that improvements here and there could raise the Prussian army to the level of effectiveness that the times now demanded. Radical change was needed, even if that implied far-reaching changes in the state and society as well. When we read their correspondence, the minutes of committee meetings, and their memoirs, we find that the officers who pressed for reform fall into two broad groups: some were prepared to accept significant changes in society and in the methods of governing the country because these alone would make radical military reform possible. Others regarded changes in the civilian realm as desirable for their own sake—they had a vision of a humane, just society in which men would not have to be forced to serve in the army but would gladly volunteer to defend their homeland. But both groups agreed on the need to cast out the old system.

It is one of the most interesting aspects of our subject that for a time, the conviction that radical innovation was needed became dominant in Prussia. Other, very different responses to Napoleon's triumph would also have been conceivable. We need think only of Austria and Bavaria after their defeats in 1805, both of which chose the very different path of moderate, partial modernization instead. Not that these are truly comparable cases—each state faced its own specific conditions—but their example does suggest that radical change was not the inevitable consequence of a major defeat. Its acceptance in Prussia even by the king and members of the old elites—often a partial, unwilling acceptance—indicates the vitality and energy that still existed in the country, even after it had collapsed before the French onslaught. The willingness to continue the war for another eight months after Jena and Auerstedt in the faint hope that further resistance would improve the political balance and create new diplomatic opportunities is another sign of this determination, a determination that alone made the reforms possible.

2

The institutional bases for reforming the army and the state were a number of royal edicts, the first of which was issued while the war was still in progress. After the armistice with France was signed in June 1807, a Military Reorganization Commission was established, which became the main engine of the military reform program.[3] The commission formed subcommittees made up of its own members and outside specialists to work on particular issues; it requested comments and suggestions from soldiers, officials, and the general public; it studied the organization and doctrine of other armies, though in the end little was adapted from foreign sources; and it worked closely with the civil administration, which was engaged in major reform efforts of its own. But the commission lacked executive authority. Its members were appointed by the king, Frederick William III, to whom it submitted periodic progress reports, and who often criticized or amended its proposals. His signature was required to turn the commission's drafts into law. It became a crucial part of the reform process to persuade and manipulate Frederick William, a man of great caution, basically conservative, and not overly self-confident. Only the shock of defeat and his practical sense enabled him to see that many of the old institutions and methods no longer worked and induced him, for a time, to support the reformers, but generally in a limiting, controlling manner. He preserved for himself a central position in the conflict between reform and the status quo, and at the end of the Napoleonic era emerged in full control of a revivified military and political system whose power had enormously increased.

Not only were the reformers forced to contend with the authority of the monarch and with conservative resistance in the army and society, they also had to deal with the French. The peace treaty signed in 1808 limited the size of the army to 42,000 men and prohibited the raising of a militia; until a sizable war indemnity was paid, French troops were to remain stationed in the country, and the reforms had to be carried out under the eyes of the occupying authority. At times the reformers could use French interference with Prussian sovereignty as a lever to persuade the king and their conservative opponents to agree to a particular change; but often the suspicions of French observers and spies merely increased conservative fears that the reformers were destroying the state and the social order.

One early measure of the reform program, however, did enjoy support across the ideological spectrum: the investigation of the conduct of every officer during the war, carried out by regimental tribunals and by a separate commission of inquiry into the actions of senior officers and the surrendering of fortresses.[4] Slightly more than 7,000 officers had served during the war, of whom some 16 percent had been killed or wounded. By 1810 the records of between 5,000 and 6,000 officers had been investigated; 208—including 17 generals—were dismissed dishonorably. This process of self-purification is another unusual episode, not only for the time, but in the history of war in general. Because senior rank and family connections proved of limited use in protecting individuals, the investigations and trials had great symbolic significance for the majority of officers who received attests of good conduct and for the army as an institution.

Of the changes carried out against this background of moral regeneration, the two most fundamental military reforms were also the two reforms that most directly and broadly affected Prussian society: one was the reform of the army's manpower policies, the other altered the methods of selecting and promoting officers. Before the War of 1806 very little ambiguity existed about who was and who was not subject to military service, even though the state's policies were based on two seemingly contradictory principles. For generations the government had asserted that, if needed, every Prussian subject had the duty to take up arms. But in fact most were legally or by tradition exempt from service. These exemptions derived from the structure of Prussian society, and it might be useful briefly to consider its salient features. We know it is misleading to think of preindustrialized society in terms of class. The character of such societies in western and central Europe is more accurately described as corporative. Corporative society was made up of many different, legally defined entities, each going far in determining the status and scope in the lives of its members. These corporative entities might be professional or craft guilds; they might be a university or a cathedral chapter; they might be a partly hereditary court of law, such as the French *parlements,* or the assembly of notables or of free peasants of a particular region; they might even be the totality of citizens of a town.

Membership in a corporative entity provided one with certain privileges, which often included exemption from military service. In 1806, for instance, every citizen of Berlin and of most other cities and towns in Prussia was exempt. The historic traditions and agreements that had led to exemption might be subject to renegotiation, of course. During the eighteenth century, the honeycomb of privileged entities making up corporative society had considerably loosened under economic and demographic pressures and under pressure from the Crown. Corporative society and modern centralized government could not easily coexist;

one limited the reach of the other, and by the end of the century, government was in the ascendant almost everywhere. One of the major and permanent achievements of the French Revolution was its destruction of corporative society in France. But as the privileges and legal inequities that it represented were destroyed, the protection against government power that many people had received from corporative society was also removed.

Until 1806 the Prussian army drew about one-third of its rank and file from foreign recruitment; the rest was provided by drafting the requisite number of men from nonexempt groups of the native population—mainly serfs, the urban unemployed, and workers who had not achieved guild status.[5] After 1807, this changed: almost all exemptions were abolished, and in 1814 the new system was logically concluded with the imposition of universal conscription. But a general obligation of this kind cut across the principle and practice of corporative society, and consequently a parallel process had to take place in the civil sphere: corporative society was destroyed, serfdom was abolished, and laws and regulations were now imposed equally on everyone, with some few exceptions. Liberal, progressive opinion, which welcomed these fundamental reforms, nevertheless objected to the abolition of military exemptions for the educated but, as in revolutionary France, had to learn that conscription was the price to be paid for legal equality.

If all Prussian males, whatever their social standing and economic condition, were now legally equal and might have to serve in the army, the code of military justice and the practice of military discipline would have to be reformed and humanized. Specifically corporal punishment had to be abolished. In the same way, if sons of the bourgeoisie could now be called to the colors, they would have to be given the opportunity to rise in the service and receive officer commissions. This brings us to the second fundamental reform, which profoundly affected the interests of the traditional elites. In 1806 slightly more than 90 percent of all Prussian officers were or called themselves noble. Bourgeois officers, the remaining 9.5 percent, served mainly in the less prestigious branches: the artillery, the engineers, and the light infantry and light cavalry. Access to commissions in the heavy infantry and heavy cavalry, and to the senior ranks, was restricted almost entirely to those of noble birth. It is true that the nobles' monopoly was not quite as firm as it appeared on paper. Many bourgeois officers passed themselves off as noble—a practice condoned more often than not—others, as they rose to higher rank, were ennobled by the state. Nevertheless the old restrictions had been a powerful bastion of privilege and could no longer be accepted. From 1808 on, anyone who passed the requisite examinations might aspire to a commission. The introduction of examinations was itself a major change. Conservatives derided it as tending to turn the army into a republic of letters, but examinations not only performed a social function by somewhat reducing the importance of family connections, they also played an essential role in the professionalization of the officer corps. Modern war demanded more diverse competence, even from junior officers, than had the wars of the *ancien régime,* in which the majority of officers had performed basic tactical tasks that required great physical courage but little thought. Now training covered more areas and became more thorough and was further institutionalized by a system of higher education for the more promising officers, culminating in a superior war college. Another side of this process of professionalization concerned the methods of advancement. The principle of seniority had governed the old Prussian army. Occasionally Frederick the Great had ignored it—especially in making senior appointments; but most officers regarded promotion by seniority as a protection against favoritism, and it certainly accorded well with the aristocratic rejection of excessive ambition and of rivalry between peers. Now seniority was abolished as a factor in the assignments of general officers and was at least somewhat weakened in the more junior grades as well. The new methods of officer selection, training, and advancement all contributed to changing the officer corps from a corporative elite to a somewhat more open, professional, and competitive institution.

Once a force of mercenaries and serfs, controlled by small elites, the army had been opened up to the whole of society. Its new, broader social composition would also make it difficult, if not impossible, to continue the old ways of fighting, the effectiveness of which was in any case open to question. Eighteenth-century linear tactics had been a highly rational exploitation of disciplined movement and massed fire; but on the march the armies were slow and cumbersome, their supply systems were too complex to be truly efficient, and the emphasis on heavy infantry and cavalry meant that reconnaissance and march security were rarely adequate. Both operationally and tactically, the need for greater flexibility and speed was apparent. From 1807 on, the army gradually developed new tactical and operational doctrines, which in 1812 were given final form by subcommittees of the Reorganization Commission. Instead of the complex rules of the old manuals, which tried to cover every conceivable eventuality, the new regulations were masterpieces of simplicity. They laid down a limited number of basic formations, posited a few general principles having to do with rapid movement and the close interaction between units, and emphasized common sense and initiative. A comparison of the bulky volume of the last infantry regulations of the old monarchy, published in 1788, with the thin infantry manual issued in 1812 will suggest the vast changes that had occurred in a few years; the difference between the two may even symbolize the reform movement as a whole.

By the late spring of 1812 when Prussia became an unwilling junior partner in Napoleon's invasion of Russia, the Prussian army—small but capable of rapid, orderly expansion—had developed into what might be regarded as the most modern military force of the time. The questions that still remained to be answered were, one, how this army would perform on campaign and in battle, and, two, whether the reformers' broader aims would eventually be realized. Would the civil and military innovations that had been hammered out with such difficulty lead to a true citizens' army serving a constitutional monarchy and a more open society, or would conservative opposition prevail and turn the achievement of the reformers to its own ends?

3

I have been describing reform as an institutional process by which the army was radically changed and have singled out such areas as manpower policies, discipline and military justice, officer selection and training, and operational doctrine. Other issues the reforms addressed cannot be discussed here: for instance, the improvement of weapons and equipment, the organization of the armed forces, the establishment of a Ministry of War, and the introduction of a new type of General Staff—an innovation that was to assume worldwide significance later in the century and remains important in our own time. But before concluding the discussion of Prussia's recovery with a few comments on the way the changes worked out in practice, I want to go beyond these impersonal factors and say something about the role of the individual in the process of reform.

Earlier I alluded to the personality and beliefs of Frederick William III. The king certainly played an important part both in making reform possible and in setting its limits. Equally influential were the personalities of the few men at the head of the reform movement, who defined and devised the specific innovations and worked to bring them to reality.

Their leader and the chairman of the Reorganization Commission was one of the most junior generals in the army, Gerhard von Scharnhorst. He had a distinguished record as a fighting soldier and was known even beyond Germany for his writings on the history, theory, and practice of war. He was also a sophisticated interpreter of the changes in warfare that had come about since the French Revolution who attributed the successes of the French armies as much to psychological and political elements as to military innovations. His theoretical and his practical gifts, rarely combined to such an extent in the same person, were given a powerful political dynamic by his shrewd management of men. Repeatedly he won over or outmaneuvered his conservative opponents, and even the king found it difficult to distrust him. As unusual as his personality was his social background. He was the son of a Hannoverian peasant, a retired sergeant, and served as a gunnery officer in the Hannoverian army until his growing reputation led to a Prussian commission and a patent of nobility. His non-noble antecedents were more or less shared by his closest associates in the reform movement, although he alone came from a peasant family. Neidhardt von Gneisenau, later Gebhard von Blücher's chief of staff, was the son of a military engineer and borrowed his title from a family with a similar name to facilitate his career in Prussia. Hermann von Boyen, who became minister of war in 1814, was the grandson of an officer who had assumed a title for the same reason. Karl Wilhelm von Grolman was the son of a bourgeois judge who had been ennobled in 1786. He left Prussia in 1809 to fight against Napoleon, first in Austria and then in Spain; in 1813 he reentered the Prussian service, rising to the rank of senior general. The most brilliant of Scharnhorst's younger assistants, Carl von Clausewitz, had both a bourgeois father and a bourgeois mother.

On the one hand, the social origins of these officers illustrate the process of co-opting carried out by the state, which needed constantly larger numbers of officers and administrators, and also by the old elites, which maintained their dominance in part by accepting able newcomers into their ranks whatever their birth. But their social origins are also a factor in the reforms. Their bourgeois backgrounds did not make these men revolutionaries but do seem to have contributed to their critical, somewhat detached view of the army and of its ties to the old established nobility. In turn, their status as newcomers certainly strengthened the conservative opposition they encountered. They saw the army as an instrument of state power, which should be subject to social considerations only as long as these did not reduce its efficiency. Some went further and wanted the army to be both an expression of a new, more open society and a means of bringing this society about. A phrase often heard at the time referred to the army as the school of the nation—"nation" here standing for a political community that no longer drew its reason for existing from political, dynastic arrangements, but from the people, their history and culture, a community in which authority no longer rested solely with the Crown. Universal military service, these reformers thought, could be justified only by a political community that represented the interests of all and in which everyone had a stake. According to this point of view, the civil equivalent of conscription was a greater measure of self-government, not only on the municipal level where it had already been introduced by the civilian reformers, but nationwide as well.

The political idealism and enthusiasm of the leaders of the military reform movement was an imponderable force, hard to measure but undoubtedly a factor in their success. As I suggested earlier, however, not every reform-minded officer shared their attitude. It soon became apparent that they formed a minority among the many officers who supported innovation. The point of view of this majority was more narrowly professional than that of their leaders. But even many of the more technical military reforms could never have been achieved by the reformers themselves. They succeeded because they were able to win over a fair number of senior and field-grade officers who had enough experience to recognize the value of innovations and enough self-assurance and professionalism to dismiss the charge that the army was being turned into a revolutionary force that would destroy the power of the Crown and of the Prussian nobility. Blücher was a good representative of this group. He trusted Scharnhorst and Gneisenau even if he did not always understand them, and he and others like him made their innovations respectable to the officer corps as a whole.

In the final campaigns against Napoleon between 1813 and 1815, the Prussian army proved to be a far more effective instrument than it had been in 1806. Not all of the credit goes to the reforms. The French could never fully recover from the Russian disaster, and Prussia instead of fighting alone was now part of a powerful alliance. Nevertheless the new army functioned more smoothly and swiftly, the supreme command had learned something from Napoleon about long-range strategic goals, and the spirit of officers and men had changed. In that connection it is interesting to note that nearly 4,000 officers who served in the War of 1806 also served in 1813. Presumably many had learned from experience, and the new institutions and methods gave greater scope than before to their energy and abilities.

But the army's successes in the field had consequences that worked against the reformers' long-range goals. Once the Napoleonic threat faded, so did the pressure to continue and complete the reform program. Even worse, some innovations could now be minimized as emergency measures whose time had passed. The most notable victim of this attitude was the Landwehr or National Guard. Between 1813 and 1815 the reformers had organized the National Guard as a separate force serving side by side with the regular army, dominated not by professional officers but by reserve officers who had returned to civilian life after active service and who represented the views of the educated and well-to-do middle range of society. Within a few years after Waterloo, the Landwehr had been so thoroughly integrated with the regular army that it lost nearly all of its special characteristics and became merely another extension of monarchical power.[6]

To sum up, when we look back at the years of Prussia's recovery after Jena and Auerstedt, we see that the army, the state, and society were reformed in tandem. But change was pushed further in the army and in the bureaucracy than in society. Conscription was not matched, after all, by constitutional government and a franchise, however limited. Society was opened up, but the old elites continued to dominate. It is striking, for instance, how quickly bourgeois officers took on the values of their noble comrades and became members of an elite that was now larger and more professional than the Frederician officer corps had been, but not all that different in its special relationship to the monarch and in its claims for political and social superiority in the country. Without external support, the liberal impulses the reformers had implanted in the service disappeared. Conservative tendencies in the state, in society, and in the army reinforced one another. The army remained an efficient institution; above all it retained the potential for further growth and modernization, and fifty or seventy-five years after the reform era it might still be regarded as the school of the nation. But what the army now taught the tens of thousands of conscripts it trained and returned to civilian life every year was to become disciplined, obedient subjects of an authoritarian state, a state that the army perhaps more than any other force in German life was turning into a world power.

3

. . . .

Britain and the
Boer War

John Gooch

The war that broke out in South Africa on 11 October 1899 came as an unwelcome surprise to the British government. Tensions between the Boer Republics and the British had mounted steadily after the Jameson raid in 1896, and during the summer of 1899, British politicians had sought to avoid a conflict. The political convictions of the Conservative administration about the necessity of resorting to force in order to secure a satisfactory settlement of the Anglo-Dutch dispute were founded in their confidence that the army was ready if necessary to fight and to win.[1] Five days before the fighting began, George Wyndham, civilian under-secretary of state for war, declared, "I believe the Army is more efficient than at any time since Waterloo."[2] The soldiers were no less confident. Viscount Garnet Wolseley, commander in chief and one of late Victorian England's military heroes, was in no doubt that the expeditionary force sent out to the cape under the command of Sir Redvers Buller would display "a very different condition of things from that which existed in the Army sent to the Crimea in 1854."[3]

The optimism and confidence of the civil and military establishments was short-lived. In the space of five days (10–15 December 1899), three generals crashed to defeat in what was immediately christened "Black Week." William Gatacre was defeated at Stormberg, largely as a consequence of inadequate reconnaissance; Algernon Methuen was worsted at Magersfontein when the Highland brigade was surprised by the unconventional tactics of the Boers, who had dug themselves in at the foot of a hill instead of at its crest as custom decreed; and Buller lost the battle of

Colenso, in good part because he lacked knowledge of the ground and of enemy dispositions.

The unexpectedness and completeness of the British defeats at the hands of an amateur force composed chiefly of farmers and one that bore little resemblance to any conventional army rattled some politicians. The defeat at Spion Kop on 24 January 1900 was a major shock to which Fleet Street reacted with anger and Whitehall with deep anxiety. To lose three battles in the space of a week could be put down to incompetent generalship; to keep losing battles thereafter suggested that the War Office and the army were prey to deep-seated deficiencies that imperiled the very safety of England itself. For the prime minister, Lord Salisbury, the problem went deeper even than that: "It is evident," he told Parliament on 30 January, "there is something in your machinery that is wrong."[4] Victorian cabinet government, though well enough designed for peaceable purposes, seemed clearly inadequate when it came to war; and since Britain faced potential enemies in Europe who were militarily infinitely more powerful than the tiny Boer Republics, Salisbury felt that the time had come to reexamine the whole structure of military policy making.

The process of inquiry into the state of the army and of the government's defense machinery, which was to be one of the most important consequences of the Boer War, began almost immediately, and soon a number of royal commissions and parliamentary committees were hard at work scrutinizing the nation's military entrails. Meanwhile, there was a war to be won. Lord Roberts, Wolseley's rival for the nation's affections, was sent out to take over command in the field from Buller with Lord Kitchener, victor of the battle of Omdurman only two years earlier, as his chief of staff. After lifting the sieges of Kimberley, Ladysmith, and Mafeking, Roberts advanced cautiously northward through the Orange Free State and into the Transvaal, capturing the Boer capital of Pretoria on 5 June 1900.

By conventional—that is to say, European—standards, the war was now over. Roberts departed for England, where he was welcomed by a grateful government with the Order of the Garter, an earldom, and a handsome sum in cash. However, the Boers saw things differently. "As everybody knows," Field Marshal Sir William Robertson remarked dryly in his memoirs, "the war lasted much longer and required far more troops than had been expected."[5] Kitchener found himself faced with a grueling guerrilla campaign in which the Boers used their mobility, field craft, and intelligence to the full. His response was to detain the civilian population in what became known as "concentration camps," to subdivide the country by building chains of blockhouses, and to hunt down the Boer commandos by sweeping the country with mobile columns. By the time the peace of Vereeniging brought the war to an end on 31 May 1902, 448,435 British and imperial soldiers had been caught up in the military effort to subdue an enemy who was never able to field more than a tenth of that number.[6]

Although the war ended in victory, neither soldiers nor politicians took any comfort in that. To contemporaries it was a special and humiliating kind of defeat. What had been envisaged as just another colonial expedition, an activity in which the British army was much practiced and commonly successful, had turned into a debacle in which a puny David had dealt the imperial Goliath some severe blows before finally succumbing to massively superior military strength that had only been generated with considerable difficulty. In the view of Leopold Amery, author of *The Times' History of the War in South Africa,* the army had been exposed as "largely a sham."[7] Many contemporaries shared this view, and it has since been echoed by more than one historian.

Nor was this the full measure of the matter. If the army had failed in particulars—and it remained to be determined exactly where and how it had failed—then it seemed no less true that the government and the state had failed in generalities. The clutch of committees that had been set up in the last decade and a half of the nineteenth century to generate and monitor defense plans and to tender military advice to the cabinet had failed to rise to a fairly humble challenge; unless they were revitalized or reformed, Britain would be incapable of defending its empire and its interests against the far more formidable threats posed by the great powers. The future structure and shape of civil-military relations stood at the head of the agenda dictated by defeat.[8]

As far as high officials in the War Office were concerned, the Boer War revealed unexpected weaknesses in a structure that had been the victim of continuous tinkering for some three decades. At issue was the extent to which the commander in chief should have authority over every branch of the army. Behind this stood the broader but no less contentious question whether soldiers or civilians should have ultimate authority over military business. The Liberals regarded the prospect of a powerful commander in chief with acute alarm as the unwanted harbinger of continental militarism. In 1895 the government had reached an uneasy compromise by giving the commander in chief authority over the so-called military departments (which meant mainly the Intelligence Branch and the Military Secretary's Office), whereas the heads of the Supply and Ordnance departments enjoyed semiautonomy.

While he was commander in chief (1895–1900), Wolseley complained bitterly, both in public and in private, about his lack of power and his inability to exercise control over the army, and later made no secret of the fact that he wanted to see a serving soldier as minister of war.[9] His successor, Roberts, soon came to share his frustration. Roberts had held office for only a very short time before complaining to the secretary of state for war about "the impossible position" he was in as "one of a board" and demanding more authority.[10] A debate in the House of Lords in March 1901 over civilian and military authority forced the government's hand. Aware that all was far from well in the War Office and smarting from defeats that could readily—though not necessarily correctly—be attributed to the weakness of the commander-in-chief's position in military counsels, the government announced an official inquiry. With that the procession of committees and commissions into matters military got under way.

The managerial revolution, which was to transform the structure and functioning of the army's upper reaches between 1901 and 1904, began with a select committee of the House of Commons which, on 8 January 1901, started inquiring into the organization of the War Office. It found the army over-centralized and yet simultaneously suffering from the long-established rivalry between the civil and military elements, and its report recommended the creation of a permanent War Office Board to combine the autonomous and often warring branches.[11] The government responded cautiously by marginally extending the commander-in-chief's authority; but the report had not been in vain. Henceforward, the debate about army reform would be informed by concepts of business management that were being consciously applied to military affairs for the first time. And in bringing the idea of a board system into the public arena, the report invited comparison between the War Office and the Admiralty, where such a system had been functioning with every apparent sign of success since the nineteenth century. Sailors crowed and soldiers fulminated as the navy was presented as a model institution that the army would do well to copy.

The War Office was not the only institution that patently needed to take a more businesslike approach toward the management of defense. The cabinet, too, required a thorough overhaul, as Lord Salisbury had intimated. Although the public lacked any firm basis for criticism until the publication of the Elgin Commission's report in the summer of 1904, the politicians were well aware that their machinery of government had been tested and been found wanting at the onset of the Boer War.

The particular target for criticism was the Standing Defence Committee of the cabinet, which had been created in 1895 to deal with military policy at the highest level but which had completely failed to generate considered and coordinated defense policies. In John Ehrman's words, "It seems to have met irregularly and seldom, and when it met to have devoted itself primarily to considering financial questions and to settling specific inter-departmental disputes." [12] Its unimportance in peacetime was more than matched by its inadequacy when war occurred. The mixture of surprise and unpreparedness that characterized the government's reactions to the outbreak of the Boer War sounded the death-knell for the Defence Committee.

Arthur Balfour, Lord Salisbury's heir apparent to the premiership, was keenly aware of its deficiencies and its limited potential. He told the incoming secretary of state for war,

> For purposes of reorganization, I believe it to be utterly valueless. We cannot, I suppose, get rid of it altogether. It may indeed perform useful work as a cabinet committee, examining, on behalf of the cabinet, schemes already more or less matured. More than this it would be folly to hope from it.[13]

Two more years were to pass, however, during which the storm of public criticism intensified and the difficulties of the Conservative government grew, before Balfour responded to direct pressure from the secretary of state for war and the first lord of the Admiralty and reconstructed the Defence Committee as the Committee of Imperial Defence, an advisory committee of the cabinet with both civil and military members. A year later, in 1903, he took over the chairmanship from the duke of Devonshire, thereby giving the committee the weight and impact in government it had hitherto lacked.[14] With that development, a new era in British defense policy making began.

The longer the war lasted, the more incontrovertible the case for reform appeared, and having once embarked on a process of official inquiry, the government was powerless to halt it. A parliamentary report—the Akers Douglas report—identified officer training as one of the causes of military weakness, castigating a system in which there was little incentive to learn and which operated on a basis of lengthy periods of idleness interrupted by brief interludes of hectic cramming. But criticism of some of the parts of the military machine could not be an acceptable substitute for an examination of the whole body military. The government could not escape mounting demands for a sweeping independent inquiry into all aspects of a military system that had apparently failed so miserably. In September 1902, against the wishes of the king who was worried about the harm it might do to the army, a royal commission was established under the chairmanship of the earl of Elgin to examine the preparations for the war and its conduct up to June 1900.

The Elgin Commission sat for fifty-five days, during which time it examined 114 civilian and military witnesses and asked them 22,000 questions. Its report, published a year later, forbore from direct condemnation and offered no clear recommendations for reconstruction, although it did find that Wolseley had failed in his offensive and defensive operations. But the evidence, in two bulky volumes, provided the army's and the government's critics with ammunition aplenty. And in a minority report, Lord Esher, a close friend of Balfour's and the king's *uomo di fiducia*, recommended the establishment of a board to run the War Office along the lines of the Board of Admiralty.

The Elgin report was delivered on 9 July 1903, but the government carefully refrained from publishing it until the end of August when Parliament had been prorogued. The time of its arrival was every bit as important as the damning nature of its contents. The cabinet was being shaken to pieces by the question of tariff reform, over which five ministers would shortly resign, while the secretary of state for war was under attack for trying to introduce a clumsy scheme intended to provide an army of three regular and three reserve corps; a scheme that had simply thrown everything into disorder without producing any substantial military force at all.[15] "The War Commission is going to give you a lot of trouble," the chief whip warned; "St. John [Brodrick] is clear of implication in matters before the war, but the evidence of his mismanagement and disregard of military opinion since the war is very strong."[16]

What the Elgin Commission's evidence made unmistakably clear was that the organization of the War Office was in complete disarray. This was due in no small part to the attempts to improve internal coordination during the war by means of an Army Board and a War Office Council. Asked about the functions of these bodies, a procession of witnesses completely contradicted one another. Sir Henry Brackenbury, director general of ordnance, said that the Army Board seldom voted, while Sir C. M. Clarke, the quartermaster general, affirmed that it did so whenever there was any difference of opinion.[17] Sir Thomas Kelly Kenny, the adjutant general, did not think that any member of the War Office Council had the right to put an item on its agenda, but Sir William Nicholson, director general of mobilization and military intelligence, declared that any member could raise any issue.[18] A senior civil servant at the War Office, Sir Guy Fleetwood Wilson, announced authoritatively that the members of the War Office Council had no right to put an item on its agenda and then had to listen while the chairman read out to him the relevant army order demonstrating that they did.[19] No one seemed to know very much at all about the Defence Committee of the cabinet. All in all, the War Office appeared to be an Augean stable awaiting its cleanser.

Whether the reform movement generated by the setbacks of the Boer War could have maintained its momentum in different political circumstances is impossible to know, although Balfour's somewhat languid reaction to the deficiencies of the Defence Committee in 1900 suggests that he might have postponed further changes in the War Office if he could have avoided them. But he could not. Lord Selborne, not only the first lord of the Admiralty but also one of the wisest of Conservative politicians, warned the prime minister that the party's hold on office depended on his grasping the nettle once more and more firmly. "The danger seems to me to lie in Army questions," he told Balfour, "and I think that unless you can satisfy the strained and irrational public sentiment of the moment about the W.O., we stand a good chance of being beaten in some division in the House."[20]

Balfour needed a new secretary of state for war, and he needed to make "a splash." Esher was the ideal man to serve his purpose; but Esher was quite unwilling to take public office.[21] So, it appeared, was almost everybody else. Half a dozen people turned him down before H. O. Arnold Forster agreed to step into what Campbell Bannerman liked to call "the kailyard." Balfour resigned himself to accepting Arnold Forster but felt that, glutton for work though he was, he could not simultaneously rebuild the army and reform the War Office. His problem was essentially a political one: how to produce "the necessary dramatic effect" without Esher.[22] The solution he devised had about it a classic Balfourian elegance. Esher would chair a committee and bring forward recommendations for the reform of the War Office along the lines of his minority report to the Elgin Commission, and Arnold Forster would implement them.

The War Office (Reconstitution) Committee produced the crowning achievements in the spasm of institutional reform that followed the Boer War.[23] Its recommendations swept away much of the pre-1899 War Office and replaced it with a functional structure designed to prepare the army for war, at whatever time and against whatever foe. The position of commander in chief was abolished and that of chief of the general staff, against which many politicians and not a few soldiers had set their faces for some fifteen years, was created; an Army Council, modelled on the Board of Admiralty, was established to bring soldiers and civilians together and to provide managerial control over the army's varied affairs; and the committee recommended that a permanent secretariat be added to the Committee of Imperial Defence to strengthen it. The Esher Committee also produced a list of candidates for the new jobs it had created and managed to get most of its favored sons installed in them. Balfour accepted all the recommendations, thereby insuring that the wave of institutional reform left in its wake not debris but a modern design for defense planning.

Asked by the Elgin Commission what "great lesson" he would draw from the Boer War, Wolseley replied: "We were not prepared for war, and are not prepared for war at all times."[24] The defeats and setbacks suffered by the army between 1899 and 1902 certainly suggested that Wolseley's view was broadly true and the military quickly set about making good some of its defects, although reforming zeal was tempered by the argument that the South African war was exceptional, even abnormal, on account of its duration, the unusual physical conditions in which it was fought (including an exceptionally clear atmosphere, which was thought to have increased the range of rifle and artillery fire), and the peculiarities of Boer tactics and organization. "It was a most extraordinary war," remarked Major General J. P. Brabazon, "fought under absolutely different conditions (it was my sixth campaign) from those of any other war I had ever seen."[25] But the process of analyzing the lessons of the war—the necessary precursor to any reforms—revealed that, although the war might in some respects have been novel, by no means had everything failed in and after 1899.

One part of the military machine went like clockwork at the outbreak of the war: the mobilization of the 30,000-man army corps Buller took with him to South Africa in October 1899. The army had Sir Henry Brackenbury to thank for this. Thirteen years earlier, alarmed at his discovery that Britain could put only one army corps into the field and then not without great difficulty, Brackenbury had started work on a mobilization scheme to remedy the defect. Plans were made and regulations issued; in August 1899 these needed only to be supplemented with tables of composition, which were rapidly prepared and distributed. When mobilization began on 7 October, the authorities in charge of the process awaited a flood of telegrams from officers who were uncertain about what to do. None came. Nor did the transportation of the newly mobilized force present any great difficulties. Close peacetime cooperation with the Admiralty, which had joined the War Office on a joint ad hoc conference on shipping on 14 April 1899, insured that the troops and their equipment were shipped to South Africa speedily and efficiently.[26]

This was not the only feather the old War Office could put in its cap. As the Elgin Commission revealed, the Intelligence Department had been remarkably accurate in its estimations of Boer strength and perceptive in its warnings about the likelihood of attack. On 11 June 1896 a report had been given to Wolseley warning that the assumption that the Boers could not advance into Natal or Cape Colony in less than the four to six weeks needed to bring in British reinforcements was no longer tenable. Over the next three years, five more warnings followed. At the same time, the Intelligence Department prepared a remarkably accurate handbook, *Military Notes on the Dutch Republics of South Africa*, which it issued first in 1898 and then, in revised form, in June 1899. It also monitored the importation of artillery, rifles, and ammunition into the Boer Republics. It underestimated Boer manpower by only 991 (out of a total of 55,641) and overestimated Boer artillery by 8 (putting Boer strength at 107 guns when the true figure was 99).[27]

The Intelligence Department had performed wonders on little more than a shoestring. On the eve of the war, its entire staff had amounted to only twenty-five, one-sixth of the number employed on the German General Staff, which had no empire to deal with; the South African section consisted of two officers and a lone clerk. The director of military intelligence, Sir John Ardagh, had used the funds at his disposal shrewdly: between 1896 and 1899 some £6,000, one-third of his total budget, had been spent gathering intelligence on South Africa.[28] However, treasury's parsimony had undeniably restricted the opportunities to prepare adequately for war: a derisory mapping budget had meant that the forces sent out to South Africa had to depend mainly on local maps, which were often well-nigh useless.

The critical failure had not been one of performance, however, but of a lack of receptivity to advice and information that should have prompted a reconsideration of policy. Wolseley carried some blame on this score, although he was frustrated by the cabinet's unwillingness to listen to him, but his lack of receptivity to intelligence was shared by other senior officers. After the war was over, Kitchener flatly refused to accept intelligence estimates of the size of the Boer forces in the field, believing them to have been ordinarily 65,000 to 70,000. This was twice the War Office estimate, and when confronted with this discrepancy Ardagh remarked that he was "inclined to think it was a very large overstatement."[29] At the back of disputes like this lay a conception of command shared by the majority of general officers that excluded any notion of interference from Whitehall with the plans and assessments made by the man on the spot and that the Boer War scarcely dented.

If Wolseley was required to bear a lot of the blame for the deficiencies in the army, he also had the right to claim a large share of the credit for its most striking success. Shocked by the administrative incompetence revealed during the Zulu War of 1878–79 and the first Boer War of 1881–82, he had revitalized the supply system, introducing a single communications authority and instituting a system of regimental transport for the first time. The quartermaster general's responsibilities were revised in 1887 to include the feeding, moving, and quartering of troops, and in 1888 Wolseley's selection for the new post, Redvers Buller, created the Army Service Corps. By the time of the Boer War, an echelon supply system had been perfected that worked well until Kitchener disbanded it and returned to column transport.[30]

Although the British military system did not completely fail when put to the test in 1899, its defects certainly outnumbered and outranked its strengths. Guns and munitions proved inadequate; prewar tactics and training turned out to be ineffective on a modern battlefield; the role of cavalry in war appeared to have drastically altered; the officers, both senior and junior, gave evidence of not being up to their jobs; and even the "queen of the battlefield" seemed to be a pretender to the throne and not the rightful monarch. "The British infantry, on whom British generals had always relied, could not shoot straight, its fire discipline was rotten, and its old-fashioned reliance on drill precluded the use of field craft and battle craft."[31] Much of the experience garnered during the war provided unambiguous evidence of inadequacies that fuelled a short but violent spell of upheaval and reform.

The Royal Artillery had not kept abreast of the artillery revolution of the 1890s and was at once outclassed by the new quick-firing guns that the Boers had bought from Krupps and Schneider-Creuzot. British field guns were outranged by those of their opponents, which could fire twice the distance and could also keep up a rate of fire two or three times their own. They also quickly ran out of ammunition thanks to serious under-estimation of the stocks that would be necessary in war. With only 300 rounds per gun and a further 200 rounds in reserve, the army's supplies were exhausted long before the middle of December 1899, and it was forced to borrow from the Royal Navy and the government of India.[32] The guns also had the wrong kind of shells. When fired from field guns, shrapnel, which formed the great bulk of artillery ammunition, merely caused the Boers to duck into their trenches; a number of officers came back from South Africa demanding howitzers and high-explosive ammu-nition or even "common shell" (solid shot filled with explosive) as the only tools with which to kill or demoralize an enemy who had taken to the earth.

The army reacted quickly to the revelation that its field guns were obsolescent. A committee set up by Roberts (a convert to quick-firing artillery since 1893) in January 1901 to reequip the Royal Artillery pro-duced agreed final designs just over three years later and put into pro-duction the 13- and 18-pounder quick-firing guns with which the British Expeditionary Force went to war in 1914. The problem of ammunition allowances was less successfully solved. A battery of quick-firing artil-lery could shoot off up to 5,600 rounds in an hour, and no horse-drawn supply system could feed such a ravenous appetite. Nor had the artil-lery used anything like that amount of ammunition: at Magersfontein, the heaviest artillery engagement of the war, British guns had fired an aver-age of 175 rounds apiece. So in 1905 stocks were set at 500 rounds per gun in the field—doubled in 1914 to 1,000 rounds—with a further 500 rounds in reserve. Requests to replace shrapnel with high-explosive were disregarded on the grounds that Lyddite had not proved to be entirely successful in South Africa; and disagreements about experience in the war also allowed the army to overlook demands for howitzers and heavy-siege artillery. By 1907 the cycle of artillery rearmament was over and thereafter government expenditure dropped.[33] The reforms had probably gone as far as was possible in the circumstances but, as it turned out in 1914, that was not far enough.

The experience of a modern battlefield, and particularly of the effects of magazine rifles and smokeless powder, came as something of a shock to some of the participants in the Boer War. "I never saw a Boer," wrote Lord Methuen after the battle of Modder River (29 November 1899), "but even at 2,000 yards when I rode a horse I had a hail of bullets round me."[34] If the experience was a novel one for Lord Methuen, it was by no means new to the British army. The effect of breech-loading rifles on the battlefield had been made plain in South Africa in 1881; and magazine rifles had demonstrated the futility of close-order and volley firing during the Tirah campaign of 1897.[35] The contrast between the battles in South Africa and the old-style infantry range at Omdurman in 1898 was a striking one, however. It brought home the need to adjust infantry tactics to the new circumstances, and a string of witnesses appeared before the Elgin Commission, with Lord Roberts at their head, to certify the need for greater individualism. On the battlefield it was hoped that this individualism would find its expression in a more open order in attack and aimed fire, instead of volley fire.

Although significant reforms in military doctrine followed quickly after the end of the war in South Africa, they were not simply the ineluctable consequence of a novel and shocking experience. Rather, the Boer War speeded up an on-going process of reform and underlined the central significance of developments whose importance had already become apparent. Some reform had, indeed, already been taking place. British infantry tactics had abandoned close formations after 1872 in favor of the "swarm" and had laid emphasis on flexibility and the accumulation of firepower at the enemy's weak points. A passage on "extended order" had appeared in *Field Exercises* in 1877, and the message had been reiterated twelve years later in *Infantry Drill*.[36] Some corps commanders had put these ideas into practice and underlined the validity of "modern" doctrine.

The battles in South Africa provided Roberts and his supporters with both the opportunity and the justification to introduce reforms they had already decided were necessary.[37] Between 1902 and 1905 something approaching a common body of written doctrine was created to replace the prewar situation in which individual commanders had exercised a gentlemanly autonomy. A small committee, which included several of Robert's favorites, quickly produced a *Manual of Combined Training* and a *Staff Manual*. At the same time, the regulations on field artillery, cavalry training, and musketry practice were revised and brought up-to-date to incorporate South African experience. Much greater stress was now laid on the power of the rifle and the effectiveness of defensive firepower.[38] However, Roberts was not persuaded by the marksmanship fetishists such as Hamilton, who wanted the British infantryman to "be able to shoot up to the standard of excellence which is required from the chamois hunter, and march 30 miles when required."[39] The volume of aimed fire laid down in an engagement was recognized to be critical; but in the last resort, quantity was more important than quality.

In general, the army accepted the reforms in tactics and training introduced by Roberts and his acolytes as a consequence of the Boer War without undue opposition. This was not the case as far as cavalry doctrine was concerned, however, and the sulfurous dispute that Roberts stirred up raged on until the outbreak of war in 1914. Discussion and debate over the role of the cavalry on the modern battlefield had begun as soon as the Franco-Prussian War was over, but the supremacy of the gun or the machine gun when ranged against horsemen armed with sword or lance was still far from obvious, and in 1899 proponents of the so-called *arme blanche* were holding their own.[40] The Boer War breathed new life into an old debate, as both sides claimed justification for their views on the grounds of concrete experience.

Roberts arrived to take up command in South Africa in January 1900 convinced as a result of soldiering in India and Afghanistan that the *arme blanche* had had its day, and in less than a week he was writing to Lord Lansdowne demanding large bodies of mounted infantry.[41] Against Boers who, in the first part of the campaign at least, never risked a charge and operated across vast spaces with mobility at a premium, what Roberts and many of his commanders wanted was men who could both ride and shoot. The enemy's tactics and the commander in chief's preferences together meant that "true" cavalry actions were few and far between—there was only one in the first four months of the war—so that when an opportunity to charge home did occur, the cavalry magnified its significance out of all proportion.

The battle of Klip Drift, on 15 February 1900, was a small-scale affair in which a few squadrons of cavalry vanquished a handful of Boers. In doing so, they successfully charged against positions manned by men armed with magazine rifles. It may well be, as has been claimed, that the cavalry "did not know how they had done it."[42] They certainly thought they knew what it meant. To Douglas Haig it seemed to be "the turning point and main incident of the war" and offered incontrovertible proof of a general truth: "In like situations of ground which is average, neither the gun nor the rifle will be able to open a road, but only the *arme blanche.*"[43]

Once back in England, Roberts set out to relegate the sword and the lance to ancillary status and to make the rifle (or carbine) the primary arm of the cavalry. His preface to *Cavalry Training*, published in 1904, did not rule out the charge but emphasized the paramountcy of the rifle for all mounted troops. The cavalry took considerable exception to this view, and a bitter controversy broke out. Over the years, both sides moved somewhat toward one another's position: Roberts came to accept that the sword had some role in battle, while French, Allenby, and other senior cavalrymen had come to believe by 1910 that the rifle would be their main arm in war. Alone of this group, Haig continued to maintain an unshaken faith in the *arme blanche;* and at less rarified levels, cavalry commanders persisted in applying shock tactics in maneuvers right up to the eve of 1914 with little apparent regard for the effects of firepower. In this respect at least, the experience of the Boer War served to confuse rather than clarify military issues.

Improved infantry training was designed to remedy the defects of the British soldier, about which there was a striking degree of unanimity in the higher reaches of the army. In South Africa, Tommy Atkins had shown himself to lack resourcefulness and to be unwilling to accept responsibility for his own well-being; content always to wait for the lead of his officers, he had shown great courage but little dedication to mastering his profession. Kitchener believed that this was largely the fault of the prewar training system and that it could be remedied, but not everyone agreed with him. Sir Thomas Kelly Kenny spoke for the conservative old guard when he declared that the raw material from which the army was fashioned was irremediably flawed:

> The British soldier we recruit as a race has very little imagination; he finds it very difficult to realize that if he does not see an enemy standing up or on the skyline he may be hiding behind a rock, this notwithstanding the training at Home, the Officers' Lectures and other means of inculcating this knowledge in the men.

The fact was, Kelly Kenny concluded, that "his mental perception is not up to requirements, nor is his education."[44]

In society at large, the view that the Boer War had revealed fundamental weaknesses in British stock generated the "National Efficiency" movement. In the army such views were the springboard for attempts to replace voluntary service with conscription. Roberts was a convinced believer in compulsion and eventually resigned from the army in order to campaign for the introduction of national conscription. Recognizing hard political realities, most of the high command accepted that peace-time conscription was impossible no matter how much they might secretly desire it. It was not until 1916, two years after a European war had begun, that the British government introduced conscription for the first time.[45]

The military advantages of conscription were that it would greatly increase the size, not only of the standing army, but also of the reserves. The War Office had scraped the bottom of the barrel to put two army corps into the field and had to depend heavily on volunteers, yeomanry, and imperial contingents to make up the numbers that were eventually required. Two successive secretaries of state for war—St. John Brodrick and Arnold Forster—tried to devise schemes to produce short service troops for home defense, long service troops to defend the empire overseas, and a reserve for wartime expansion. Both failed ignominiously. During their struggles, another royal commission reported that of the auxiliary forces the militia were unfit to take the field and the volunteers not qualified to face regular troops. Eventually, between 1906 and 1908, Richard Haldane rebuilt the army as six divisions backed by a Special Reserve and the Territorial Force. Haldane's army was perhaps the most important outcome of the Boer War, but it was an indirect legacy for it was built on the mistakes of his predecessors.[46]

The Boer War also revealed deficiencies in both the quality and the quantity of the officer corps. According to Roberts, the proportion of failures among commanding officers and brigadiers during the campaign was "considerably larger" than among the junior ranks; this was a view Kitchener shared. But the junior officers of the regular army showed a distressing lack of professionalism, which raised questions about their education that were borne out by the Akers Douglas report mentioned earlier.

A large part of the problem of inadequate commanders revolved around the acute shortage of staff officers. A trickle of trained officers had graduated from the Staff College during the 1890s, and although they sufficed for an army of 90,000, their numbers were totally inadequate for a force of a quarter of a million and could not be supplemented, since there were few officers with staff experience to fall back on. Moreover, the importance of a trained staff had been underlined by complexities of the field—a scale of operations with which the army was completely unfamiliar. Against a handful of Boers the problem had not proved fatal, but it might easily be otherwise. "If we take the field with a force the size of this one against an European enemy," wrote Sir James Grierson in July 1900, "and continue in our present happy-go-lucky style of staffing we shall come to most awful grief."[47]

Part of the solution to the problem was to update and improve the Staff College curriculum. Sir Henry Rawlinson, another of Roberts's protégés, was brought in as commandant in 1903 and over the next three years he reformed its program, reducing the number and importance of examinations, increasing the emphasis on continuous assessment, and revising and modernizing the course content to include study first of the Boer War and later of the Russo-Japanese War.[48] Important though these reforms were, however, they depended for their ultimate success on solving the broader problem of the British army. The Esher Committee began the process, creating a general staff and choosing its first leaders. Unfortunately some of their selections proved disastrous—notably the founding chief of the general staff, Sir Neville Lyttleton, whose interests largely revolved around cricket and lawn tennis. Gradually matters improved as the infant general staff played an increasingly important role in the work of the Committee of Imperial Defence; but the young plant had only just taken root when war broke out in 1914.[49]

If something could be done about the staff, it proved very much harder to do anything about the run-of-the-mill officer, since the army was entirely in the grip of its main supplier of officer candidates—the public schools. From 1873 onwards, the War Office had tried to raise the level of scientific knowledge of its schoolboy entrants and to substitute written and colloquial French and German for the classics but had encountered unyielding resistance. In 1893 it had capitulated, recommending that the entrance examinations conform to the public school syllabuses. Latin therefore remained compulsory.

The Boer War—and the recommendations of the Akers-Douglas Committee that English, mathematics, and French or German should become compulsory subjects in the army entrance examinations and Latin and Greek only optional ones—prompted the army to try once more to bring the public schools into line with its requirements. These proposals unleashed a storm of opposition, in which the *Times* took the lead; and the Headmasters' Conference argued that the principle of nonspecialization in public service examinations must be adhered to at all costs. What this amounted to, as the headmasters of Eton and Wellington both forcefully pointed out, was that Latin should be a compulsory subject qualifying a young man for a commission.

The pressures for reform were sufficiently strong for the War Office to alter this requirement for compulsory Latin; but Latin and Greek remained optional and the public schools' grip on officer education did not loosen. The director of military operations wrote in 1907:

> I wish we could make more rapid progress in obtaining a thoroughly modern education for our officers. I should like to see classics entirely eliminated from all our Army examinations but the public schools and older universities, which can teach nothing else, will try and force us to accept their obsolete ideals as long as they possibly can.[50]

However, some senior officers continued to hold firmly to the view that a good public school education was all that the aspiring officer needed. "No other nation," declared Douglas Haig in 1906, "is in possession of the class from which we *ought* to recruit our Officers and have in the past recruited them from until we went cracky over the 'exam. system.' "[51]

If the public schools and older universities were unhelpful in improving the intellectual qualifications of intending officers, they proved crucial in solving another of the problems revealed by the Boer War. The army was quite unprepared for the large demand for junior officers (in the eighteen months from January 1900 they had to find more than 3,000 in excess of normal demand) and had been forced to turn to the militia, with results that were by no means satisfactory.[52] Officer training could be considerably speeded up if a part of it were provided in the schools and universities, and a proposal that all boys over the age of fifteen should be given instruction in drill, maneuver, and the use of arms was made at the Headmasters' Conference in February 1900. Lord Lansdowne, the secretary of state for war, rejected this idea on the grounds of expense and the likely opposition it would provoke, but the proposal gave rise to an intense public debate.

The indirect effects of this debate were considerable. In 1906, a War Office committee (the Ward Committee) used the ideas that had been debated a few years before as the basis for their proposal to establish an Officers' Training Corps (OTC) in two divisions: a junior division in the schools and a senior division in the universities. Success in a test and possession of a certificate would give four months' exemption in each division against a total requirement of twelve months for a reserve commission. The OTC was duly established in March 1908 and rapidly attracted support from the schools and universities: by 1911 there were 152 junior and 20 senior contingents and 24,000 officer cadets. The system triumphantly proved its worth during the First World War when the OTCs provided approximately 100,000 of the 230,000 officer candidates.[53]

The Boer War was a watershed in British military history. The army's defeats on the South African veldt and the inquest into their causes and significance fuelled a wave of reform that, over the next half dozen years, changed many of its weapons, many of its tactics, most of its organization, and all of its command structure. One measure of Britain's achievement in recovering so quickly and so successfully lies in her military performance in the First World War. Had the lessons of the South African War not been learned in time, the war in France and Flanders might have been decided much more quickly and much less favorably for Britain. That was certainly Sir William Robertson's view, and more than one historian has echoed his claim that the Boer War changed everything—or at least everything that subsequently mattered.[54]

Change there certainly was after 1902, but measuring it is a difficult matter, not least because of the difficulty of creating exact and agreed criteria by which to operate. Thus, one historian has seen half a century of conscientious unprofessionalism followed, after 1899, by a novel spirit of professionalism and a "pathetic emphasis . . . upon scientific soldiering" where another has seen the persistence across a supposedly crucial divide of "the amateur, traditional ideal of war."[55] In this respect, the consequences of the Boer War would appear to have been partial and incomplete at best; but both these adverse judgments turn on definitions of amateurism and professionalism that are at least debatable. All in all, both appear harsh.

Important though the Boer War undoubtedly was in the subsequent development of the British army, it should not be viewed in artificial isolation from what preceded and what followed it. When the war began, the army had already experienced a first wave of modernization that had borne fruit in the mobilization schemes, in new principles of training, and in supply and transport. All of this was of great importance but not enough to ready the military machine for the kind of war that then took place. As the historian of these reforms has pointed out, "just as army reform was adjusted to embrace the warfare of the Victorian age, this was coming to an end."[56] Nevertheless, the achievements of the Victorian army paved the way for the changes made by its Edwardian successor.

Less than a year after the end of the war in South Africa another war began. When viewed in the context of British military reform, the Russo-Japanese War was something of a mixed blessing. British observers—of whom the lion's share were with the Japanese—used the opportunity and authority of their position to preach doctrines to which they were already converted. The "lessons" of this war, and particularly the idea that success in battle did not depend on firearms but that victory was "actually won by the bayonet," were used to refocus the army's tactics on a vision of future war that departed in several vital respects from South African experience.[57] Books were rewritten to assist in this process of readjustment. In 1907 a new edition of Sir Edward Hamley's classic textbook *The Operations of War Explained and Illustrated* was prepared with revisions by one of Haig's protégés. The section on tactics was replaced with a new one dealing chiefly with the Russo-Japanese War in which there was only a single brief reference to the Boer War—contradicting those such as Roberts who had argued that "future armies must consist of mounted men."[58] Happily there was no Laurence Kiggell to undo the reforms in defense bureaucracy that the Boer War had initiated and that were perhaps the most important outcome of the war on the veldt.

4

Caporetto
Causes, Recovery, and Consequences

Brian R. Sullivan

The reality of the Battle of Caporetto has become lost in
its mythical reputation. For most people who have even a superficial
knowledge of military history, Caporetto serves as the classical example
of mass cowardice and military incompetence. Since the Great War, the
very name Caporetto has served to blacken the reputation of Italian arms.
Even among those educated persons with little or no interest in the events
of the First World War, Hemingway's widely translated *A Farewell to Arms*
has created a vivid impression of the battle as a nightmare of panic and
cruelty from which the novel's protagonist is wise to flee. It seems to matter little that Hemingway's recreation of the battle was entirely imaginary.
Contrary to the novelist's later contentions, he spent October–December
1917 working as a newspaper reporter in Kansas City.[1]

Ignored or forgotten in this misrepresentation of the historical record
are the equal or greater disasters suffered by other participants in the First
World War. Caporetto was hardly the nadir of military performance in
that conflict. The crushing defeat and flight of the French First, Second,
and Fifth Armies in August 1914, the rout of the Russians at Tannenberg
that same month and at the Masurian Lakes a month later, the simulta-
neous destruction of much of four Austro-Hungarian armies in Galicia
and the flight of their survivors, the collapse of the Russians in Poland
and their panicked withdrawal following the Gorlice-Tarnow offensive in
May 1915, the rout of the entire Romanian army in October–December

The author expresses his gratitude to his friend, Lucio Ceva, whose valuable and perceptive comments on an earlier version of this chapter led to its great
improvement.

1916, the collapse of the Nivelle Offensive and the mutiny of nearly half the divisions in the French army in April–May 1917, and the precipitous retreat of the British Fifth Army in March 1918, not to mention the disintegration of the entire Russian army in 1917, all these come readily to mind as other "Caporettos" or worse.[2]

Nonetheless, there is no doubt that their defeat in October–November 1917 proved a terrible setback for the Italians. But the extent of their collapse is not nearly as remarkable as the degree of their recovery over the next few months. That story and its consequences form the subject of this chapter.

On 24 October 1917, the Austro-Hungarian and German forces on the Italian front began the seventeen-day offensive that would become known as the Battle of Caporetto. This battle is generally considered the greatest military defeat in Italian history.

Exactly one year later, on 24 October 1918, the Italian army began the offensive operations that would later receive the name of the Battle of Vittorio Veneto. The fighting culminated ten days later in the complete collapse of the Austro-Hungarian army. Of all the victorious powers in the First World War, only Italy ended the conflict with a total military victory on the battlefield over its opponent.[3]

Considering the Italians' failure to recover from other defeats during wartime, such as those suffered in 1849, 1866, 1896, and 1943, this series of events represents a remarkable anomaly in Italian history. When one considers the magnitude of the defeat at Caporetto and the fragile nature of Italian society in the early twentieth century, the victory of Vittorio Veneto seems all the more remarkable.

In order to appreciate fully the Italian recovery from Caporetto, we need to consider both the reasons for the defeat and the dimensions of Italian losses. As is the case for most military catastrophes, the causes of Caporetto were both military, in a narrow sense, and psychological-political. The latter, however, were probably more important. In fact, the relatively successful attempts to correct the nonmaterial weaknesses of the Italian army and to bolster public support for the war by emergency political measures offer the major explanations for Italian recovery from defeat.[4]

To a large degree, it was not the Italian nation but the political and military system it had inherited from the kingdom of Sardinia that suffered defeat in October–November 1917. Contrary to patriotic rhetoric, then and since, Italy had not been united during the Risorgimento by a widespread popular uprising against the foreign oppressors, the Austrians, and the local Italian tyrants they supported. True, Garibaldi's volunteers and the followers they attracted in the south did play a large role in liberating the kingdom of Naples. But it was the army of the kingdom of Sardinia (commonly known as the kingdom of Piedmont) that conquered and united most of the Italian peninsula. Although there is no precise statistical data to prove the thesis, it now seems evident that the majority of the Italian population during the Risorgimento wanted freedom, not unification under the House of Savoy. Many considered the northerners as their conquerors, not their liberators, after the reality of life under Piedmontese law became clear. A very bloody guerrilla insurrection in the former kingdom of Naples, from 1861 to 1866, a conflict that claimed far more lives than all the other wars of the Risorgimento, offers proof that many southern Italians strongly resented their northern Italian invaders.[5]

This insurrection received considerable support from the Papacy. The Piedmontese had invaded and annexed much of the Papal States in 1860 and conquered the remainder in 1870, save for the small enclave of the Vatican. The kingdom of Italy was proclaimed in the spring of 1861, but at first, it represented an enlarged Piedmont, which imposed its monarchy, its constitution, civil service, police and armed forces on the rest of the new state. This reality was symbolized by the decision by both Cavour, the last Piedmontese and the first Italian prime minister, and his king, Vittorio Emanuele, to retain his old title. The first king of Italy was officially Vittorio Emanuele *II*.[6]

This state of affairs was gradually transformed by a growing sense of national unity over the next half century. Nonetheless, the Italian government, the aristocracy, and the officer corps remained dominated by Piedmontese. Many non-Piedmontese were co-opted into the ruling class. Even fifty years after national unification, however, many Italians still felt that they lived under the rule of their northern Italian conquerors. In addition, the Papacy refused to recognize the legitimacy of the kingdom of Italy, creating a permanent crisis in church-state relations, in a country in which the vast majority of the population was at least nominally Catholic.

Added to these divisions were the social and political conflicts caused by terrible national poverty. In 1915, Italy remained a predominantly agricultural society in which the majority of Italian peasants were landless farm laborers. Italy had enjoyed a remarkable spurt of industrial development in the twenty years before the First World War. But this had occurred so rapidly that social and political consciousness had not caught up with economic reality. In addition, as in the case of early industrialization elsewhere, this produced a wretchedly poor and exploited working class that viewed their bosses, the government, the police, and, all too often, the army, as partners in their oppression. In 1912–14, the government finally established universal manhood suffrage. But the strength demonstrated by the new Catholic and the older socialist parties showed just how many Italians felt totally alienated from the groups that had ruled Italy since 1861.[7]

To make matters worse, Italy entered the First World War in May 1915 against the will of both the majority of the population and the Parliament. This is not to argue that no Italians wanted to fight on the side of the Allies. Many did. However, most Italians still preferred neutrality. But the king came to see war against Austria-Hungary as in Italy's best interest.

The Piedmontese Constitution of 1848, under which Italy continued to be governed, gave the monarchy extraordinary powers in foreign policy and military matters. Since the days of Cavour, the prime minister had acquired many of these powers. However, when in 1914–15 the king, prime minister, foreign minister, and colonial minister formed a conspiracy of four in order to bring Italy into the conflict, Parliament was unwilling to prevent war at the cost of forcing the king's abdication. Under these unpromising circumstances, Italy entered the conflict. Even if not displaying active hostility, the majority of the population regarded the war with a dangerous sense of indifference.[8]

As had been the case for the other belligerents in the summer of 1914, the conspirators and the Italian Army General Staff expected a short war along mid-nineteenth-century lines. After the events of August 1914–May 1915, however, this attitude amounted to criminally willful ignorance. Italian military attachés in Paris and Berlin had sent detailed, accurate warnings to Rome about the nature of the war, both before and after Italian intervention. Both the conspirators and most on the General Staff, particularly army chief of staff, Luigi Cadorna, refused to heed these warnings.

Cadorna deserves no special blame for persisting in the notion that conventional frontal attacks could pierce field fortifications. That critical error was shared by most commanders in the 1914–17 period. But Cadorna's insistence that Italy's war would be a short one and that his planned attacks could succeed without adequate artillery preparation represent errors that stain his reputation irredeemably.[9]

Given national poverty and the low level of Italian industrialization, the Italian army had long been woefully underequipped. Furthermore, the recent Libyan War of 1911–12 had exhausted the army's stocks of supplies and ammunition. As a result of these shortcomings, the army entered combat in the spring of 1915 with neither the training and tactical doctrine nor the machine guns and artillery to engage successfully in trench warfare.

The situation could have been even worse, however. To his credit, General Alfredo Dallolio, head of the army's artillery and engineer directorate, had anticipated a long war and Italy's participation in it as early as August 1914. Throughout the period of Italian neutrality, Dallolio had created mechanisms for close coordination between the army and industry. Soon after intervention, Dallolio persuaded the government to order a full-scale "industrial mobilization." In October 1915, Dallolio was appointed high commissioner and, shortly after, undersecretary for arms and munitions, granting him greater control over war production. But many months were to pass before Dallolio's work resulted in a significantly increased flow of artillery, shells, and heavy equipment to the army.[10]

Meanwhile, the army had to struggle with the consequences of insufficient arms and supplies. Infantry training was particularly affected. For most recruits it consisted only of close-order drill, the firing of half a dozen rounds from their bolt-action rifles, and learning to advance by bounds in long parallel lines. Not even when assigned to combat units did soldiers receive additional instruction prior to entering the trenches, except in rare cases of initiative shown by a few responsible commanders. The training of artillerymen was better, for that branch had long been the pride of the Italian army, carrying on the old Piedmontese tradition. But poor communications equipment and lack of combined arms training for the infantry and artillery frequently led to disaster on the battlefield. Furthermore, until 1917, the artillery lacked sufficient shells and suffered throughout the War from confused and inconsistent operational and tactical doctrine. The artillery often lifted its fire either too soon during an assault, leaving the attacking infantry exposed without fire support, or too late, resulting in casualties by friendly fire.[11]

In addition, there were the problems presented by the terrain of the battlefield. Unlike on the western front, the two sides on the Italian front fought not in mud but on a waterless, stone-covered plateau or on the slopes of the Alps. The resultant increase in fragmentation caused by exploding artillery shells produced 70 percent more casualties per detonation than on other fronts. Worsening this problem for the Italians was a severe shortage of doctors, medical supplies, motor transport for the wounded, and an overall low level of sanitation among the troops. The result was not only a terrible casualty rate but also the outbreak of epidemics of cholera and typhus throughout the Italian army in July–August 1915.[12]

General Cadorna's attitudes and personality proved even more detrimental to the effectiveness of the Italian army. Cadorna epitomized the traditions of the old Piedmontese army. He rejected any political interference in his operations, literally assuming dictatorial powers in the war zone and refusing to communicate his plans to the government. In this, he received the full support of the king.

Cadorna possessed a strong personality and an inflexible sense of duty. He insisted on assigning his only son to a combat unit at the outbreak of the war, behavior hardly typical of other Italian generals. Basically a shy man, Cadorna felt grave discomfort in the company of strangers and lacked even a trace of the common touch. He also suffered to a degree from a persecution complex and reacted to both criticism and unsolicited advice with deep suspicion. As a result, he exhibited little interest in the morale of his soldiers or the opinions of his field officers. As he saw it, their duty was to fight and die. His duty was to direct the army.

Cadorna showed virtually no concern for the multitude of factors, ranging from proper training to adequate transport, that translate plans on paper into success on the battlefield. If plans did not succeed, Cadorna believed, this was the fault of the weaklings or cowards who failed to execute them properly, not of the man who had conceived them. If an army fought long and hard enough, he believed, the enemy would collapse. Simply put, Cadorna was a general with an eighteenth-century mentality directing a twentieth-century war. One could make similar comments about Douglas Haig and other ranking military leaders in the Great War. But British, French, and German field commanders enjoyed access to more resources than their Italian counterpart, making it easier for them to survive the consequences of their mistakes.[13]

Despite his commander's shortcomings, the average Italian soldier displayed remarkable loyalty for the first two years of the war. Because of the low level of education and skills among the population, the army gradually adopted policies that gave exemptions to industrial workers and concentrated the literate in such branches as the artillery, motor transport, and the engineers. The result was that the infantry, who suffered the overwhelming majority of the casualties, was mostly composed of southern Italian peasants, often illiterate. Many of these men had led lives of such deprivation that army service actually offered them a higher standard of living than they had previously enjoyed.

Particularly after mid-1916 the improvement of army services—the regular availability of meat, fish, coffee, wine, and liquor; of warm clothing, bedding, and shelter; of rudimentary medical services and occasional entertainment; of pay and mail delivery—helped to prevent rebellion against the horror of life in the trenches. As in all armies, however, rear-area troops enjoyed far more comfort than the combat forces. The latter rarely received their full quota of rations and equipment.

Furthermore, these southern peasants were, in general, originally so apolitical that Catholic and socialist opposition to the war did not initially affect them, while the traditional nature of their lives had prepared them for the blind obedience that the officer corps demanded. As the war dragged on, however, many soldiers became literate, while general exposure to men from other regions with greater political consciousness provided another form of education for the typical infantryman. We might consider the Italian army of 1917 to have been composed of a remarkably unsophisticated group of common soldiers. But compared to the army of 1915, it had made enormous strides in understanding the nature of the war and the shortcomings of the Italian high command under Cadorna's rigid leadership. By mid-1917, the rank and file had become increasingly restive.[14]

Details about the conduct of the war between May 1915 and October 1917 must be found elsewhere. Suffice it to say that the Italian high command and officer corps showed as little understanding about how to break the deadlock of attrition warfare as did the leaders of the French and British armies. In addition, the problems created by the nature of the battlefield and the initially low output of the Italian arms industry made the likelihood of a breakthrough seem even more improbable on the Italian front. It is all the more remarkable, therefore, that by the late summer of 1917, the Italians were on the verge of smashing their way through the Austro-Hungarian defenses and knocking the Hapsburg monarchy out of the war.[15]

How had this happened? First, because the Austro-Hungarians did not present the same formidable opponent to the Italians that the Germans did to the Allies on the western front. The Austro-Hungarians had suffered very heavy casualties among their officer and noncommissioned officer (NCO) corps in the opening battles of the war against the Russians in 1914. Their army never recovered from this blow. After Italy entered the struggle, the Hapsburg army had to conduct a three-front war until the collapse of Serbia in late 1915 and a two-front war thereafter. Austro-Hungarian arms production proved markedly insufficient, the level of training and leadership in the Imperial and Royal Army remained considerably lower than in the German army, and severe morale and discipline problems existed in some of the Hapsburg army's regiments due to ethnic conflicts.[16]

The terrain on the Italian front, however, still gave the Austro-Hungarian defenders an enormous advantage. Yet the Italian infantry, by demonstrating an extraordinary courage and willingness to attack, overcame this advantage. In some cases, these qualities had been promoted by ruthless military discipline involving numerous executions by firing squad, the frequent decimation of units that showed hesitation under fire, and the widespread sackings of officers of all ranks.[17]

Nonetheless, the Italians had carried out eleven great offensives along the southeastern portion of their front by August 1917, a feat without precedent on any other front of the Great War. They had exhausted themselves in the process. But the Austro-Hungarian high command knew that one more Italian offensive meant the probable collapse of the Hapsburg army. That was why the Austrian Kaiser Karl begged the Germans for assistance in late August and why the Germans agreed to send seven divisions to the Italian front to participate in the offensive that would become known as the Battle of Caporetto.[18]

Meanwhile, grave problems had come to afflict the Italian army, as well as the nation itself. True, by mid-1917, many of the material problems that had afflicted the army in the previous years had been overcome. Firepower had been greatly increased. There had been considerable improvement in training, in tactics, in medical care, and in the proficiency of the high command. But morale at home and at the front had dangerously weakened. War weariness grew ever heavier. The antiwar pronouncements of the Papacy and the socialists had finally begun to seriously affect the common soldier and the average citizen.

In July, during a speech in the Chamber of Deputies, the prominent socialist leader, Claudio Treves, voiced his hopes: "next winter, no more in the trenches!" The following month, in a diplomatic note to all the belligerent powers, Pope Benedict XV described the war as a "useless slaughter." This latter pronouncement particularly alarmed Cadorna, for at that moment his troops were locked in what he hoped would be a breakthrough offensive on the Bainsizza plateau.[19]

Throughout the home front, the passive acceptance of the war had gradually turned into bitter resentment. During 1916 and 1917, the inflation rate in Italy rose considerably higher than in Britain or France. Price controls and rationing were poorly administered, and social discontent gradually worsened. In August 1917, strikes by industrial workers in Turin over food shortages quickly developed into violent demonstrations against the war. Many participants saw the ongoing Russian Revolution as a model for Italian emulation and made their feelings clear. The government responded by dispatching troops to quell the worker violence with brutal force. Striking munitions workers were deprived of their exemptions from military service, drafted, and sent to the front. By unfortunate coincidence, some of the units formed from these workers were assigned to the Caporetto sector.[20]

Despite their actual success in the battle on the Bainsizza, the terrible cost of pushing the Austro-Hungarians to the very brink of defeat convinced many Italian soldiers that the struggle was futile. Since May 1915, the army had advanced a maximum of only eighteen miles from the prewar frontier at the cost of 900,000 casualties. The high command had promised victory with the great offensive of August–September 1917 and the army had given everything in that series of attacks. When the Austro-Hungarians did not collapse, the heart went out of the Italian army. By the eve of Caporetto, many units in the Italian army were in psychological crisis. But some generals paid little attention.[21]

A detailed explanation of the technical military reasons for the German-led success against the Italians in October–November 1917 would require a book in itself. But a number of factors should be mentioned. By mid-1917, Austro-Hungarian military intelligence had succeeded in breaking all Italian army codes, gaining as a result a tremendous operational advantage on the eve of Caporetto. The arrival on the Italian front of sizable German forces brought new infiltration tactics, a superior use of artillery, flamethrowers and, especially, new poison gasses (against which Italian gas masks offered no protection). Furthermore, the failure of most Italian commanders to employ defense-in-depth and coordinate artillery fire with defensive plans contributed to the German breakthrough. But perhaps the most important factor was the timing of the Austro-German attack, begun less than two months after the collapse of the Italian August–September 1917 offensive. The impact of Caporetto might be compared to what the French army would have suffered if the Germans had been able to launch their spring 1918 offensive in the summer of 1917 instead, shortly after the Nivelle offensive and the French army mutinies of May 1917.[22]

During the first days of the battle, many Italian units fought stubbornly and effectively. In other cases, however, panic took hold. Officers, even a few generals, abandoned their men and unit cohesion collapsed. Such incidents led to an unraveling of the entire front held by the Italian Second Army. After it began, even the most severe disciplinary measures failed to halt the rout. Cadorna then had no choice but to order the retreat of all of his forces. The disintegration of huge sections of the Italian army that followed took the Germans and the Austro-Hungarians completely by surprise. They were not prepared to take advantage of the opportunities that developed after their breakthrough, partially a reflection of the perennial German difficulty in translating operational success into strategic victory. Yet the German and Austro-Hungarian commanders should not be faulted too severely for failing to anticipate the consequences of their breakthrough. After all, the origins of the Caporetto catastrophe lay at least as much in Italian history as in German military skill.[23]

Figures help to tell the story. On the eve of Caporetto, the Italian army had sixty-nine divisions in the field, armed with 6,900 cannon, 2,400 mortars, and 8,700 machine guns. By the time the army dug in on the western bank of the Piave River and halted its retreat on 9 November, it had shrunk to thirty-three divisions with 3,800 cannon, 700 mortars, and 5,700 machine guns. The most telling figures are the losses in manpower: 40,000 killed or wounded but at least 280,000 prisoners and 350,000 deserters or stragglers. The army had not so much been smashed on the battlefield as fragmented during the retreat. In essence, the willingness of many Italian soldiers to go on fighting and dying for the monarchy of the kingdom of Sardinia, for the Piedmontese officer corps, for the benefit of a nation state that had meant only high taxes and hopeless war had come to an end.[24]

Given the dimensions of their victory and the legendary flexibility of the German army staff system, why did the Germans and Austro-Hungarians fail to knock the shattered Italian army out of the war altogether? For one thing, the Italian army that survived the retreat from Caporetto was obviously the half of the army still disciplined and willing to fight. In addition, Cadorna had displayed his worst, but also his best, qualities during the retreat. On the one hand, in a dispatch to the king and the cabinet on 28 October—soon made public—Cadorna had blamed what he called the cowardice of his own soldiers for his defeat. This truly monstrous slander, which deeply angered the king and which Cadorna himself later sincerely regretted, would cost the general his command.[25]

Yet, Cadorna had generally kept his head during the withdrawal. After failing to hold the line at the Tagliamento River, he had arranged the remnants of his army in secure defensive positions along the Piave, anchored to the north on the massif of Monte Grappa. When he was relieved and replaced by the Neapolitan Armando Diaz, as chief of staff on 9 November, Cadorna turned over an army that, for all its problems, held a front line reduced from 650 to 400 kilometers. As the Italians had retreated, the Allies had rushed eleven British and French divisions to Italy. These forces remained on the Mincio River, however, some seventy-five miles to the west, awaiting developments. When the enemy resumed the attack on 10 November, the Italians faced fifty divisions and 4,500 guns ranged against their thirty-three divisions and 3,800 cannon.[26]

Meanwhile, on 8 November, Vittorio Emanuele III had met at Peschiera with the British, French, and Italian prime ministers. Representatives of the Italian high command had made a very bad impression on their allies a few days earlier, at an emergency conference held at Rapallo. As a result, the British and French had demanded a meeting with the king. The unhappy Vittorio Emanuele III reluctantly admitted that his army suffered from many weaknesses and, on Lloyd George's inquiry, announced Cadorna's impending replacement by Diaz. But the king stressed his deepest conviction that, whatever its previous and present shortcomings, the Italian army would defend the line of Monte Grappa and the Piave at all costs. The monarch's expression of faith in his army, enunciated in his fluent English and French, proved decisive in bolstering Allied support for Italy. But Diaz was one of the few Italian generals who actually shared his sovereign's belief in the continued possibility of victory.[27]

Almost miraculously, the fighting remnants of the Italian army lived up to the king's expectations. Two days after the Peschiera Conference, the Germans and Austro-Hungarians began a series of attacks across the Piave and up the slopes of Monte Grappa. These continued with little interruption until 4 December. After a brief respite, they renewed their attacks from 11 to 19 December and, again, from 25 to 30 December. The Italians drove off all these attacks.

Only during the first week of December, by which time it was fairly clear that the Italians would hold, had the Allied forces entered the line. During the same period, the Germans withdrew three of their seven divisions from Italy and the remainder followed over the next few weeks. By New Year 1918, the forces on the Italian front had shifted to a rough balance. Italian self-confidence had been severely wounded, however. Months would pass before the Italian high command and government acquired something approaching an objective view of the Italian military situation.

The basic cause of the defeat of the Central Powers' forces was stubborn Italian resistance, reinforcements in the form of virtually untrained seventeen- and eighteen-year-old Italian conscripts, and the exhaustion of the attackers. It should be stressed, however, that throughout the second half of November 1917, the Italian infantry were often reduced to fighting with literally no more than rifle butts, bayonets, rocks, and sometimes their bare hands. The lack of ammunition, even for those heavy weapons that had survived the retreat from Caporetto, meant that the defenders frequently fought with absolutely no fire support. Under these circumstances, the small arms, machine guns, artillery, and ammunition supplied the Italians by their allies in the period December 1917–January 1918 proved of crucial value.[28]

Encouraged by this defensive victory, the new Italian high command decided to launch a limited counteroffensive operation in late January 1918. The purpose was both psychological—to demonstrate to the nation, the army, and the enemy the extent of Italian recovery—and practical—to recapture key terrain that represented the deepest Austro-Hungarian advance during the fighting over the Christmas period. The result was the so-called Three Mountains offensive of 27 to 31 January. At the cost of heavy fighting and rather severe losses, elements of two Italian divisions succeeded in seizing all their objectives, inflicting heavy casualties on the Austro-Hungarians, and capturing several thousand prisoners in the process. The massive employment of artillery (900 guns and heavy mortars on a front only three kilometers long), including a heavy use of mustard gas shells, showed that the high command had already learned many lessons since Caporetto.[29]

But this victorious defense and limited counteroffensive did not so much represent recovery from defeat as the limits placed on the scope of the German and Austro-Hungarian victory. By late January 1918, the Italian army had lost one great battle yet had rebounded to severely check its previously triumphant enemies. More important, the Italian army had managed to avoid losing a great war by a heroic last-ditch defense and an extraordinary psychological recovery. To move beyond this limited success toward victory would require a herculean national effort.

Early steps had been taken in November 1917–January 1918 by the round-up of deserters and stragglers. On 2 November, the high command ordered every soldier to report to his unit within five days or risk the death penalty. It appears that 141 deserters were executed for failing to comply but many thousands of others escaped. On 10 December, a royal decree gave deserters nineteen days to report to duty or face the firing squad thereafter. Before the deadline, 27,000 men responded to this order. In the Modena region south of the Po, two huge camps were established for the collection and reorganization of 200,000 infantrymen and 80,000 artillerymen who had become separated from their units. These troops were reformed into 104 infantry and 22 field artillery regiments. Another 13,000 combat engineer troops were gathered into a third camp nearby, outside Reggio Emilia. A stricter reinterpretation of the grounds for medical exemptions added 150,000 recruits to the army. Finally, additional eighteen-year-olds of the conscript class of 1900 were called to duty several months early in mid-March 1918.[30]

In the meantime, the Italian government had seriously considered seeking a separate peace with the Central Powers and would continue to do so for several months. Initial secret contacts with the Austro-Hungarians revealed that Italy could gain peace without losing territory. But the Italians would be excluded from any influence in the Balkans and the Near East and would have to accept a subservient status in a Europe dominated by Germany. Later talks suggested that Italy might even gain some territory in exchange for peace with an increasingly desperate Austria-Hungary. But these rewards seemed very meager compared to the losses that Italy had already suffered in the fighting. Faced with these alternatives, the government felt it should remain in the war.[31]

Partly on the basis of the idea that three heads are better than one, partly to dilute the power that Cadorna had held, Diaz had been assigned two deputy chiefs of staff, Generals Gaetano Giardino and Pietro Badoglio.Particularly in November 1917, the three generals showed certain weaknesses in their new exercise of command. Although more flexible mentally, intellectually subtle, and politically astute than Cadorna, Diaz lacked his predecessor's adamantine will. But his deputies gradually came to make up for Diaz's shortcomings. The new chief of staff concentrated on strategy and political liaison with the government, Giardino

turned his attention to tactics, operations, and logistics, and Badoglio devoted his energy to rebuilding army morale. In February 1918, Badoglio
managed to obtain Giardino's exile to Versailles and to find a substitute
for him in Badoglio's own protégé, the able staff officer Colonel Ugo
Cavallero. By then, the Italian army had begun to regain its strength.[32]

But full recovery from Caporetto required many months and began
something of a revolution in Italian society. Most essential was a change
in Italian consciousness that transformed perceptions of the conflict from
just another dynastic war (although on an unprecedented scale) into a
national struggle against a foreign invader. In other words, in late 1917,
for the first time in Italian history, a sense of genuine patriotism began
to take hold within most segments of Italian society. The danger to all
Italians represented by an invasion of the northeastern edge of the Po
Valley helped to settle the question of the meaning or importance of the
war. Even most socialists rallied to the cause, if only temporarily. Rather
suddenly, the majority of Italians found themselves supporting the war
effort, whatever their feelings about how Italy had entered the war.[33]

This consensus, combined with American and Allied willingness to
supply Italy with loans, food, coal, and raw materials, allowed the Italians to rebuild their army and the high command to restructure the army
both materially and psychologically. Reversing Cadorna's policies, the
high command and government began to coordinate their activities. Both
agreed that the army would need many months of recuperation before
it could undertake any major offensive actions. Furthermore, the flow of
Allied supplies to Italy by sea remained seriously restricted by U-boat
attacks until June 1918 and this hindered the material recovery of the
Italian army until the summer of that year.[34]

Nonetheless, a wide range of new services and reforms brought radical changes to the trenches beginning in early 1918. These included the
opportunity to take leave on a regular basis; propaganda directed at improving morale; far greater attention paid to the personal comfort, nutri-
tion, and health of the common soldier; the opening of officer candidate
courses to qualified NCOs; and bringing officers with combat experience
into staff positions.

Improvements in combat efficiency were introduced, including
greatly improving the firepower of companies and battalions; granting the
commanders of platoons, companies, and battalions far more autonomy
in battle; introducing more effective coordination and communication
between artillery and infantry units; creating new assault units modeled

on the German storm troops (*Arditi*) and aviation ground attack units to support the infantry. In general, a new tactical doctrine emerged that stressed firepower over the use of massed manpower, both in the defense and the attack. As a result of these improvements, the British and French were able to withdraw six divisions from Italy to face the German spring 1918 offensive on the western front, while the Italians sent two divisions of their own to France.[35]

Under Dallolio, recently promoted to minister for arms and munitions, a massive reorganization of industry allowed the Italians to produce over 5,000 artillery pieces in 1918, as well as 6,500 aircraft (compared to 5,500 from May 1915 to December 1917), and to raise artillery shell production to 90,000 a day. By June 1918, the army had been re-expanded to fifty-four divisions and by October to fifty-seven divisions, along with three in Albania, a double-sized division in Macedonia, and two in France. In addition, the army had formed twenty-three independent battalion-sized *Arditi* units on the Italian front and four more on other fronts. In other words, the Italians had finally harnessed all the material and psychological forces of the nation to the war effort. They had learned how to wage a modern total war.[36]

The most important factor in recovery, however, was the generation of enthusiasm for the war both in the army and in the general population. It was not enough to create motivation through fear of the consequences of a victory by the Central Powers. Instead, the government decided to promote the vision of a new Italy that would arise after victory. Fundamental to this promise was that of a new economic order in which the returning landless peasant veterans would receive their own farms, in which worker veterans would find decent jobs with ample salaries and social welfare benefits, and discharged junior officers and NCOs could obtain university degrees or technical training at government expense. These promises did succeed in rousing real national support for the war effort on a sustained basis. But these were promises that the government could later ignore only at its peril.[37]

However, Diaz feared that the Germans and Austro-Hungarians would launch an offensive on the Italian front before his army had enjoyed sufficient time to recuperate. In March and April 1918, he braced himself for an enemy attack launched in coordination with the German spring offensive in France. No such blow landed—although Ludendorff had seriously considered it. Instead, the first test of Diaz's rebuilding efforts came in June. Caporetto had granted the Austro-Hungarians a six-month reprieve from collapse. But both the success of the German spring offensive and the ethnic fracturing of the Dual monarchy counseled a final effort to knock Italy out of the war.[38]

Although originally scheduled for late May, the Austro-Hungarians delayed their advance across the Piave River until June 15. By then, deserters had already alerted the Italians to the coming offensive (although not to its precise timing). Furthermore, Diaz had learned many lessons since Caporetto and Italian defenses proved effective. The Italians had constructed fortifications in depth, properly sited their artillery, and discovered the advantages of counterbattery fire and counterattacks. Against the Austro-Hungarian onslaught, the Italian artillery brought to bear no less than 7,000 artillery pieces which fired some 3.5 million shells. In addition, Diaz brought large numbers of aircraft into his defensive scheme. He employed bombers to attack enemy concentrations on either side of the Piave and used fighters to strafe the pontoon bridges which the Austro-Hungarians had thrown across the river.

Meanwhile, severe food shortages had brought many Austro-Hungarian units close to starvation. Exhausted by their attempts to overcome the Italian defenses, thwarted by the rain-swollen river, and unable to reach the Italian food supplies they had been promised as the fruits of success, Austro-Hungarian morale eventually cracked. After eight bloody days, the Italians achieved a complete, if defensive, victory. After visiting the front, Kaiser Karl ordered the evacuation of the bridgeheads. The last offensive in the history of the Austro-Hungarian army resulted in 140,000 casualties.[39]

The repulse of the Austro-Hungarian June offensive, followed over the next weeks by incontrovertible evidence that the German offensives in France were failing, produced ever-growing political fissures within the Hapsburg monarchy. Food shortages in the cities and in the army made daily life increasingly hard to bear. Military discipline began to collapse. In September, the Bulgarians collapsed and sued for peace. Allied forces began a sweep up through the Balkans toward the southern frontier of Hungary.[40]

Nonetheless, Diaz and his advisors had decided not to launch a major counteroffensive in 1918. The army had exhausted its ammunition supplies in its defensive victory in June. More important, grave doubts remained about the new Italian army's willingness to undertake large-scale offensive operations. The high command decided to prepare for a major attack in the spring of 1919. Despite growing pressure from the Allies, seeking even indirect relief from German attacks in France, Diaz rejected their pleas for action from late June through July. Beginning in August, however, the military situation in France reversed dramatically and the Allies began to drive the Germans back.

In late September, the Italian high command began studying plans for an all-out offensive. From early October, Diaz came under increasing pressure from his government, the French, and the British to take action. Finally, realizing that a German collapse might be imminent and that the war might end with Italy in a highly unfavorable diplomatic position, Diaz agreed to launch a major offensive late that month. While Cavallero and Badoglio later received credit for planning the operation, its true originator was the reserve Major Ferruccio Parri.[41]

But the rising waters of the Piave convinced Diaz to postpone the attack. Instead, he granted permission to the recently recalled Giardino, who had been appointed commander of the Italian Fourth Army, to make an improvised assault on the enemy-held side of Monte Grappa. This desperate plan seemed better than complete inactivity. The attack began on the first anniversary of the German breakthrough at Caporetto. For two days, however, Giardino's troops failed to gain ground, due to poor infantry-artillery coordination. They lost 15,000 dead and 20,000 wounded in the process and Diaz suspended operations. But the attacks had caused the commitment of some of the meager reserves of the Austro-Hungarians.

Meanwhile, a growing number of other Austro-Hungarian reserve units mutinied and refused to move up to the front. News of the fracture of the empire into national states, which had begun in mid-October, had reached the troops. Realizing the opportunity these events had created, Diaz then ordered a general attack across the Piave according to Parri's original operational concept.

Despite the growing disintegration of their empire, the Austro-Hungarian army heroically resisted Italian assaults until the evening of 29 October. Heavy rain, which had turned the river into a torrent, greatly assisted the Austro-Hungarians. Open revolution had broken out across the empire, however. Once the Italians and their British and French allies had gained firm bridgeheads across the Piave, Austro-Hungarian resistance finally began to crack. Early on 30 October, the Austro-Hungarian army began to retreat. This withdrawal soon dissolved into a rout, especially after Hungarian, Czech, and Croat units rebelled and began marching for home. Already, the Austro-Hungarian high command had requested an armistice. By the time this had been arranged late on 3 November, perhaps 250,000 prisoners, 5,000 cannon, and 2,900 mortars had fallen into Italian hands. The battle left some 9,900 Italians dead and 25,800 wounded.

Taking the name from a key town captured in the advance, Diaz officially designated the victory the Battle of Vittorio Veneto. By the time the armistice on the western front took effect on the morning of 11 November, the Italian army had reached the Brenner Pass and was preparing to advance into Bavaria. Italy had won a great victory, but counting those who would later die from wounds or disease, the price was 709,000 military dead and another 100,000 civilian fatalities.[42]

From the Italian point of view, this might seem like a happy ending to an otherwise tragic story. Such was not the case, however. Just as the causes of the defeat at Caporetto had deeper roots than simple military weaknesses, the recovery from Caporetto involved more than the victory of Vittorio Veneto. The consequences of the recovery proved catastrophic for the Italians because they helped lead to twenty years of fascism and to Italy's disastrous involvement in the Second World War.

All European countries involved in the Great War, victors or vanquished, emerged from the struggle seriously wounded in one way or another. Italy, initially weaker than most of the other combatants, came out of the war in a perilous state. In particular, the Italian political system had suffered grave damage. Despite the manner in which the monarchy, high command, and the government managed to rally the nation in the period after November 1917, the old power structure never recovered the prestige it had lost at Caporetto. To a large degree, these ruling groups remained on trial, pending the outcome of the Paris Peace Conference and the reshaping of postwar Italian society. When the Italian representatives at Versailles failed to acquire both the European and colonial territories that most Italians believed were Italy's due, national confidence in the government greatly declined.[43]

In the meantime, with the end of the war, Allied and American loans and supplies of grain and raw materials ceased. Rather quickly, the Italian economy fell into a depression, further aggravated by soaring inflation. As a result, the successive postwar governments proved unable to redeem the promises made during the war, nor even to ensure political and labor peace. The old order had been deemed a hopeless failure. The new seemed little better.

On the other hand, those groups that had opposed the war, principally the socialists, the new Communist party, and the Catholics, failed to offer attractive alternatives to most Italians. In the first postwar election of November 1919, these groups did gain a majority in Parliament. But the Papacy forbade the Catholics to form an alliance with the officially atheistic socialists. Soon, many voters grew disillusioned with parties that could not come to power, however attractive their promises sounded. Furthermore, in the eyes of other Italians, these groups had been discredited by Vittorio Veneto. Many moderate or conservative Italians (and these comprised the majority of the voters) saw the political opposition as tainted by treason and defeatism. In addition, the specter of revolution on the Bolshevik model, while hailed by the Left, terrified many Italians. At the same time, the Catholics seemed to threaten a hidden return to papal rule by a church that had clearly favored Austria-Hungary over Italy during the war.[44]

Although it took until early 1920 for these political realities to coalesce, when they did, one group stood ready to take best advantage of the situation: the new fascist movement of Benito Mussolini. From the founding of the movement in early 1919 until its collapse in the summer of 1943, the fascist core was formed by veterans of the First World War, especially the veterans of the *Arditi* assault units and many of the former NCOs and junior officers who had become the backbone of the army after Caporetto.

These men seized on the widespread impression that both the old ruling class and the new Left had equally betrayed Italy. They argued that the sense of national unity forged after Caporetto could be restored and institutionalized by a form of permanent national military discipline. Such a spirit, the fascists argued, could make Italy invincible, allow the nation to regain what the politicians and diplomats had thrown away after Vittorio Veneto, and propel Italy to world power status. Using an unprecedented combination of violence and lies, these men seized power over the state in late 1922 and established their fascist dictatorship little more than two years later.[45]

This is not to argue that Italian recovery from Caporetto led directly to the establishment of Mussolini's dictatorship nor that the fascist regime was Italy's inevitable fate. In seeking recovery from defeat, however, the Italian government of 1917–18 had unleashed powerful forces that eventually turned against it in the postwar period. The failure to channel these forces in constructive directions did lead to fascism.

The Italian government had not been foolish to remain in the war after Caporetto, nor to seek a victorious conclusion to the conflict. The government's major mistake was made in May 1915: the irresponsible decision by a few men to force the Italian people into the war against their will. Thereafter, whether it won or lost the conflict, the Italian nation was bound to undergo strains beyond the capacity of the traditional social order and system of government to sustain. Caporetto was not so much the revelation of the hopeless inadequacy of the old order as it was the final shudder of a system that had committed suicide through war. Thereafter, any recovery could not mean a revival of the old but only the birth of the new. Unfortunately, due to the mistakes of 1919–22, it meant the birth of a monster.

5

. . . .

From Politiques en Képi
to Military Technocrats
De Gaulle and the Recovery
of the French Army after
Indochina and Algeria

Martin S. Alexander and Philip C. F. Bankwitz

Between 1946 and 1962, in the jungles of Indochina and in the desert djebel of Algeria, the French army waged two wars. It lost each of them. More seriously, perhaps, in the process of losing these wars the French army—and in particular its senior officer corps—also lost its sense of mission and duty, its military professionalism and discipline, and its subordination to civilian control.[1] In consequence, the task confronting Charles de Gaulle after his return to power in May 1958 was a daunting one. His challenge was to reorient an army—and more particularly an officer corps—that had lost its way. The requirement was rendered all the more delicate by the fact that de Gaulle himself was, arguably, that army's most famous and influential *indiscipliné* of all.[2]

The issue facing de Gaulle was that of reimposing a willingness to obey on an army officer corps that had almost entirely lost the habit of obeying. The task was all the more difficult because of the example of insubordination to civilian governmental authority that de Gaulle himself had set in June 1940, when the French army had disintegrated as a coherent fighting force and much of France lay under the occupying heel of the German Wehrmacht.[3] But the context of the Fall of France mattered less than the fact that the precedent had been established in 1940 for autonomous political action by the generals. At Bordeaux, on 15 and 16 June 1940, the military had decisively entered the arena of French politics. Defeatist or at least fatalistic generals, with supreme commander Maxime Weygand to the fore, drove the divided civilian government into a corner. They thereby facilitated the political takeover of the *capitulards* who were ready to seek accommodation with Hitler.[4]

De Gaulle proclaimed that France had lost, not the war as a whole, but merely its first battle. On 18 June 1940, he appealed from London over the British Broadcasting Corporation for the French people to defy the new collaborationist regime of the army's luminary and 1914–18 hero, Marshal Philippe Pétain, and to rally to his side. Yet, for all de Gaulle's courage and vision during modern France's darkest hour, his rebellious action in proclaiming himself an alternative locus of political authority and a personification of "a certain idea of France" nevertheless cast him as a central figure in the modern French army's descent into politics.

After his return to power—via the aptly code-named Operation Resurrection of 13 May 1958—it was essential for de Gaulle to restore an older, but severely weakened, tradition of military apoliticism.[5] To ensure the success of his own regime and implement his own political agenda after 1958, he needed to resubordinate the French officer corps to civilian political control. It was a steep gradient he had to climb, for he bore significant responsibility for enabling the officers to break out of their purely professional role during the earlier crisis.

Yet de Gaulle was extraordinarily successful in accomplishing the task he set for himself. By May 1968 the army's loyalty—and, more important, its obedience—had ceased to be an issue. Despite the turbulent *événements* which shook de Gaulle's presidency that spring and which played a part in the general's resignation from office in 1969 and in his death in 1970, the army stayed off the political stage.[6] In fewer than ten years, de Gaulle had engineered the recovery of the army and, indeed, of the armed forces as a whole. He had healed the twin traumas of defeat and demoralization that had resulted from the lost colonial wars of 1946–62. This chapter will explore the ways in which de Gaulle performed this remarkable and enduring feat.

After more than a decade of debilitating colonial warfare, French officers in general—and those of the army in particular—were disillusioned and discontented in 1958. Those in the more senior ranks, the generals and colonels, knew little except defeat. The heroic days of the Liberation of 1944–45 had faded into a dimly remembered past. The glamorous dash of General Philippe Leclerc's Second Armored Division leading Allied tanks into Paris on 25 August 1944 had assumed an almost mythic status; the triumphal progress of Marshal Jean de Lattre de Tassigny's First Army to the Rhine and Danube was almost forgotten.[7] Looking back over twenty-five years, the general pattern of French military performance seemed inglorious, ignominious, and shamefully unsuccessful. From the failure to react to Hitler's reoccupation of the Rhineland on 7 March 1936, until the Evian Accords brought a cease-fire to the conflict in Algeria on 19 March 1962, the chronicle of French military undertakings was littered with broken dreams and shattered illusions, especially the unending, morale-sapping cat-and-mouse war of attrition in Indochina.

And it was surely there, in the steaming jungles of the Far East, that the French army officers' loyalty to the Republic slowly but inexorably evaporated.[8] Metaphorically at least, Indochina was the grave of France's two most distinguished and lionized leaders of the "new army" of 1944–45—Leclerc and de Lattre—and the real grave of their sons, Henri de Hautecloque and Bernard de Lattre, the last-named an only child. Inspector-general of the French army by 1947, Leclerc's unexpected and accidental death in a plane crash in the Sahara that year dealt a body blow to the veteran cadres, noncommissioned officers, and field commanders. It was to Leclerc that this backbone of the postwar French army looked for inspiration as they set off to swell the Corps expéditionnaire d'Indochine, and it was to them that the counterinsurgency war against Ho Chi Minh and the Viet Minh's chief of staff, General Vo Nguyen Giap, was entrusted.[9]

Near-mystical powers were associated with Leclerc and his Second Armored Division, and it was widely expected that the success which had come the way of French arms in la métropole under his leadership in 1944–45 would be replicated in Indochina. "Hier Strasbourg . . . demain Saigon" (Yesterday Strasbourg . . . tomorrow Saigon), as one recruitment poster for the French expeditionary corps to the Far East characteristically put it—a French tank and its young commander in the foreground against images of the Strasbourg cathedral and the temple of Angkor Wat. "Without the Empire," the black deputy from Guyana Monnerville had declared before the Consultative Assembly on 25 May 1945, "France would be just another liberated country. Thanks to the Empire, France is a victorious nation." [10]

Leclerc's unforeseeable and shocking removal had the frightening force of oriental fatalism. Whether or not Leclerc fell victim to French communist sabotage of the aircraft taking him on his inspection (the machine crashed on takeoff in a Saharan sandstorm on 28 November 1947), his death seemed to be an omen of impending disaster.[11]

De Lattre's death from cancer in 1952 similarly assumed symbolic significance.[12] His period of command in Indochina in 1950 and 1951 had witnessed the application of new French methods against the Viet Minh, especially his policy of taking the war to the enemy more vigorously. The background to de Lattre's assumption of command in Indochina could hardly have been less auspicious: in an eight-day action of unprecedented savagery during October 1950, Giap had launched a Viet Minh force of 25,000 men which overwhelmed several French battalions engaged in withdrawing from exposed outposts at Cao Bang and Dong Khe on the Chinese-Vietnam border—a rout that then precipitated an unforced abandonment of the Tonkin rear base at Lang Son. This defeat, quickly known among the French as "the disaster on RC [Route Coloniale] 4," sent a shudder of alarm through the entire colonial community.[13]

Given the fireman's job after this flare-up of communist operations, de Lattre arrived in Saigon on 17 December 1950. Two days later he inspected a vast review of French forces at Hanoi, with the object of restoring his troops' morale. Breathing fresh inspiration into the French forces of the Orient, de Lattre told his men that there would be "no more giving up of ground; from now on you're going to be under command." [14] Throughout 1951 he ordered and implemented aggressive search-and-destroy operations; employed heavily defended combat "boxes"; and instituted vastly improved methods of command and control, air-support, artillery fire plans, and deployment of intelligence assets.[15] Ultimately, the enhanced French performance on the ground during this phase of the conflict in northern Indochina proved illusory. For every French victory, the enemy always appeared capable of fighting another battle and, indeed, as hindsight has made obvious, the tide was perceptibly turning in favor of Giap's Viet Minh guerrillas even before de Lattre's final illness. "From July 1951 onwards," as General Raoul Salan, de Lattre's second-in-command and later immediate successor, subsequently conceded, "guerilla operations resumed a primary importance and the mobile war once again reverted to a secondary activity." [16] In the spring of 1953, barely a year after de Lattre's death, Giap widened the war by invading Laos and took the Viet Minh offensive into the Mekong Delta.

Psychologically, however, the French military's greater effectiveness during 1951 had contributed to a crucial attitudinal change in the politico-strategic perspective of the French officer corps. The change was apparent not only in field commanders but also among staffs. De Lattre restored self-confidence, an aura of competent generalship, and a will to win. Hiding his terminal illness from his troops and staff, de Lattre handed over command in Indochina to Salan and left Haiphong for Paris on 20 November 1951. "By the imposition of his will," as one of his biographers has concluded, "he had retrieved a situation that even the most sanguine had regarded as beyond repair . . . he had restored the will and the power to achieve victory on the battlefield, and abroad he had restored a measure of confidence in French arms. With unrivalled eloquence . . . he had pleaded a cause that even in his own country was hardly more popular than in America."[17]

Yet this very success contained within it the seeds of much of the malaise that would afflict the French army officer corps over the following ten years, until de Gaulle's concession of Algerian independence on 1 July 1962. With disastrous long-term political and psychological consequences, the de Lattre era convinced many mid-rank and senior French officers that the war in Indochina was one they could have won. Unwittingly, but crucially, de Lattre thus assumed a key part in the emergence of a conspiracy theory, of a peculiarly French version of a stab in the back.

According to this version, France lost at Dien Bien Phu in May 1954—and thus lost possession of Indochina at the ensuing Geneva peace talks—not because of bad generalship, mistaken doctrine, and inadequate operational and tactical methods. Instead, French officers persuaded themselves, France lost as a result of entirely extraneous factors—factors quite unconnected to the political context and ideological character of the war, to the French military performance, to the mismatch between policy ends and military means, and to the ultimately "unwinnable" nature of the conflict.[18]

A myth about France's defeat in Indochina came gradually to permeate the officer corps as the 1950s wore on. Defeat was a product of misfortune or downright bad luck, of desertion by France's allies, and of the faintheartedness of her civilian leaders 8,000 miles away in Paris. On the first count, the soldiers cited the deaths of Leclerc, and especially of de Lattre. Sentimentally but nonetheless faithfully reflecting this view of the officer corps was the remark made in 1975 by General Salan, that in 1951 de Lattre "happily combined civil as well as military powers [in his one person] over the course of this crucial year." De Lattre's "shining willpower, his diplomatic skills and his proverbial brutality . . . succeeded, in the blink of an eye, in transforming French as well as Vietnamese hearts and minds." For Salan, de Lattre's death in January 1952 was accompanied by the "disappearance of that terrible energy and the permeation of clear and lucid thinking that guided our teams towards the political and military goals that were essential."[19]

On the second count—the insinuation of desertion by France's allies

—the officers of the Corps expéditionnaire d'Indochine lamented the restraints and conditional cooperation imposed by Presidents Truman and Eisenhower of the United States. The French were frustrated, mystified, and angered by administrations in Washington that spoke of Cold War, the domino theory, and their wish to crusade in defense of western liberal values, while in practice offering little military assistance and much anti-imperialist criticism of French efforts in Indochina. Twenty years later, a senior commander such as Salan would concede that the most "radical" strategic options, such as the U.S. atomic bombing of Viet Minh rear bases, which some French officers had requested, were "unthinkable," even as a way to break the "strategic impasse" that had descended upon the war by 1953. Yet at the time of Dien Bien Phu, no such objective recognition of the unfeasibility of these desperate fantasies was apparent among the French colonial officers.[20]

On the final and most significant count, French officers in the field condemned the lackluster, pusillanimous attitude of the civilian authorities to the war in the Far East—a war that the troops in the theater of operations had been compelled to wage with one hand tied, metaphorically, behind their backs, and a war about which most French civilians knew little and cared even less.

Here the myth of the undefeated French army in Indochina acquired a further and more striking dimension. The army that had confronted Ho and Giap over the course of eight years was a professional army. It was the army of France, but not the army of the French people. Many of the French units that served in the Far East were themselves composed of manpower recruited from Indochina and from other French dependent territories. Others, especially units of the Foreign Legion, carried on their rolls a great number of mercenaries of German, Belgian, Latvian, Lithuanian, and, above all, French origin whose adopted identities obscured Second World War records that were at best murky when not downright murderous.[21]

Of the white French troops, the majority were long-service career professionals. They included the Régiments d'Infanterie de la Marine, the Régiments de Char Coloniaux, the Foreign Legion, paratroop regiments, elements of the air force as well as the navy, and special forces. They did not include the draftees, the conscript units formed from the young men of metropolitan France obliged to fulfill a minimal requirement of compulsory military service. Thus the French forces that fought in Indochina remained socially as well as professionally isolated. Inevitably, as is generally the case with armed forces based on professional service, comparatively few French families had sons, husbands, and fathers serving in the Orient. Few were directly touched by the casualties. For the vast majority of the French voters, citizens, and taxpayers, the war in Indochina was not their war. It was a distant conflict whose costs to France seemed of questionable benefit and whose politics were incomprehensible. The officers and men of the French Indochina Expeditionary Corps gradually acquired the jaundiced outlook on the French people back home that had marked British Field Marshal William Slim's Fourteenth Army in Burma in 1943–45. Like Slim's men before them, the Corps expéditionnaire d'Indochine felt themselves to be a "Forgotten Army."

As the euphoria of the de Lattre era faded, so the jaundice gave way to bitterness, anger, and recriminations. The antagonistic sentiments of those soldiering in Indochina hardened toward the politicians in Paris with the advent of Generals Raoul Salan and Henri Navarre, conservative in political sympathy and unconnected with the Gaullist armies of the 1945 Liberation. Defeat at Dien Bien Phu brought the process of disaffection to a climax: the army felt it had been let down.[22] Many of its officers began to speak openly and to write under pseudonyms in semiofficial publications, such as the *Revue de défense nationale*, of a "betrayal." After the cease-fire negotiated at Geneva in June 1954, it was "onto the political authority, or more precisely onto the dearth of this authority, that each and every one of the Indochina combatants laid the responsibility for an eight-year-long struggle." Navarre, for his part, did not shrink from articulating a stab-in-the-back theory, arguing that "the mistakes . . . have been too numerous and too continuous to be attributable merely to men or even the governments who have been in power. They are the fruit of the regime itself. They stem from the very nature of the French political system."[23] The same officers came home in 1954—in many cases only to be dispatched across the seas once more scarcely three months later. This time, however, their destination lay only three hundred miles away on the shores of North Africa, where the French army was asked to confront a second nationalist insurrection, this one with much more menacing consequences for France.

Pierre Mendès-France, who formed a government in June 1954, pledged to extricate France from Indochina in thirty days or resign. At the Geneva Conference in June 1954, he redeemed his pledge. More than this, Mendès succeeded in advancing the cause of independence in two of the constituent territories of Afrique Française du Nord—Morocco and Tunisia, both of which emerged as fully sovereign nations in 1956. This much was a remarkable and enlightened accomplishment for a government whose tenure of office lasted barely seven months, from June 1954 to January 1955. The Mendès-France government appeared so fresh, so accomplished, so energetic because it worked against a deadening backdrop of what has been termed a "*société bloquée*," a "stalemated society." [24] Not surprisingly, since Mendès's government took on the brightness of a seven-month-long shaft of light, the "Mendès experiment" would in due course be elaborated into a myth in which Mendès was depicted as the savior manqué of the Fourth Republic.

An audit of Mendès's stewardship of the French army, however, presents a far less favorable outcome. For one thing, as Herbert Tint has observed, had "there not been a good deal of difference between the approaches of the French army and the population of metropolitan France to the problem of Indochina, it is doubtful if Pierre Mendès-France could have made peace with the Indochinese even in 1954." It was above all "the lack of interest, and in the end the distaste, of the French for the Indo-Chinese war [that] helped to make the Geneva agreements of July 1954 possible." [25]

Furthermore, Mendès belied his reputation for innovative and imaginative solutions, for cutting political Gordian knots, by responding in an entirely orthodox manner to the nationalist Moslem insurrection that erupted in Algeria in November 1954. Along with most of his contemporaries, Mendès subscribed unquestioningly to the conventional wisdom—indeed the constitutional fact—that Algeria was not a colony or overseas territory, but an integral part of France. France, after all, encompassed the *métropole* and Algeria from Dunkerque to Tamanrasset. This was a position Mendès did not renounce until May 1956, when he resigned from the Mollet government for ceding to demands from the French settlers for harsher repression of the Algerian Moslems.[26]

Administratively, Algeria returned deputies to the National Assembly in Paris elected from the three constituencies of Oran, Algiers, and Constantine. A million and a half white Algerians of French descent, the *pied-noirs*, lived in the country. Quite literally, these French Algerians had been "born with sand on their shoes." Metaphorical though it was intended to be, the aphorism contained an eloquent summary of pied-noir politics, prejudices, and self-definition. It spoke for the pied-noir community's intransigent opposition to any change in the status of

Algérie française. Organizationally, the pied-noirs were a vocal, tightly knit caucus, politically and journalistically well connected. For them, as well as for their relatives and supporters in metropolitan France, there was no alternative but to meet the insurgency of the Moslem Algerian Front de Libération Nationale (FLN) of Ferhat Abbas with crushing military force.[27]

If the attitudinal and psychological baggage carried into the war in Algeria by French veterans returning from Indochina was influential, so too was the fallout from the frustrated Anglo-French intervention at Suez in November 1956. The nationalist Egypt of President Gamal Abdul Nasser provided resources and shelter for the FLN's political leadership. A blow at Nasser was regarded by the French military high command as a blow at the "rear base" of the Algerian nationalist cause. For the Suez intervention, some of the best French army and air force units, including Massu's Tenth Parachute Division, were temporarily withdrawn from the conflict in Algeria and given special training for the drop on the canal zone. When the operation was torpedoed at the instigation of the Americans, the frustration and fury of French officers was unbounded. "The disappointment of the Suez operation," notes one scholar, "was directly proportionate to the enthusiasm which it had aroused [among French officers]."[28] For General Maurice Challe, a coup leader in favor of Algérie française in April 1961, "the West's loss of Africa began on 6 November 1956 at Suez."[29] And as the historian Herbert Tint has remarked: "The fact that the French army thought the politicians in Paris were responsible for robbing it of victory at Suez, just as the same politicians had previously been accused of having abandoned it in Indo-China," militated against French soldiers' tolerating compromise negotiations with the FLN "without first having secured a military victory."[30]

Consequently, the war in Algeria widened in 1956–57. At times it was an urban guerrilla war, and it engaged the French army in operations against a wily, slippery, and often-invisible enemy whose strongholds and solace lay in the souks and casbahs of the cities, in Bône, Philippeville, Constantine, Oran, and Algiers itself. Yet it was, simultaneously, a war of the wide-open spaces, a struggle of hide-and-seek, and of short but violent pitched battles in the djebels to the south of the coastal cities and in the desert dunes of the Central Sahara. The war in Algeria was also a mountain war, conducted in the inhospitable crags and passes of the High Atlas and the Aurès. It was a war of high technology, of helicopters and missiles. But it was also a low-intensity war, one demanding counterinsurgency techniques, one in which the operations took the form of small raids and ambushes—a conflict where the premium lay with the side most adept at camouflage, concealment, cunning, and surprise.[31]

But, above all, it was a war that the French waged with a different army from the one that had sunk into the quagmire of the Indochinese paddy fields and jungles. In Algeria, in a sense, the French met the people's war with the people's war. The irregular *fellaghas* of the FLN were Algerian patriots. They were fighting for their vision of a free and sovereign Moslem Algeria. But the French fight to retain Algérie française was every bit as much a popular struggle, in the proper meaning of the term. For the pied-noirs the war was a bid to preserve their way of life, their very existence. French citizens though they were, they were not people of metropolitan France.

This war rapidly assumed a higher level of brutality than the one France had recently lost in the Far East. For one thing, the officers were angry, their thresholds of restraint lowered by their determination to achieve victory. The French officer corps in Algeria—and especially the hardened veterans of the Foreign Legion and paratroops—traded in the coin of ruthlessness and violence. The "French army's bitterness at its defeat in Indo-China led it," according to Tint, "to blame its own ineptitude on the regime of the Fourth Republic which had allegedly starved it of supplies."[32] The career officers thirsted to avenge themselves on the Algerian FLN. No longer would the French army fight with one hand tied behind its back.

From the outset, this was quite openly acknowledged to be a *sale guerre*—a "dirty war." No holds were barred, no quarter was given. Torture of captured FLN soldiers and of suspected FLN sympathizers became routine. "Search-and-destroy" or "cleansing operations" (*opérations de nettoyage*) were regularly mounted into the casbahs and souks of the cities and towns. Especially uninhibited were the methods applied by the elite parachutist regiments of the Foreign Legion. Most notorious of all was the Tenth Paratroop Division commanded by one of Leclerc's erstwhile "young lions" from the Second Armored Division of 1944–45, General Jacques Massu. Under Massu's stern direction, this division relentlessly pursued and eradicated the FLN's urban strong-points and cells in a series of operations that lasted for nine months of 1957 and became known as the "Battle of Algiers."[33] The French relied heavily on the considerable support they received in intelligence, infiltration, and active operations from the pro-French Moslem Algerians, the *Harkis*. The FLN despised the *Harkis* even more than they did the French or the pied-noirs, for they regarded them as treacherous collaborators with an occupying colonial power.[34]

There were two further distinguishing features of the French army's war in Algeria. The first concerned the composition of the military contingents with which France waged the war. In Algeria, in contrast to the campaign in Indochina, the war was conducted largely by a conscript force. Because Algeria was constitutionally a part of France itself, there was no legal impediment to deploying the annual conscript contingents to North Africa. And it was a socialist-led government, that of Guy Mollet, which in 1956 "decided to call up the reserves and commit the conscript army to the struggle." The resultant military buildup took French force levels in Algeria from some 50,000 soldiers in November 1954, to just under 200,000 men when Mollet assumed office in late January 1956, to 500,000 by that year's close.[35] As Mollet put it, expounding his thinking in April 1956 in connection with the deployment of the conscripts, "The action for Algeria will be effective only with the confident support of the entire nation, with its total commitment."[36] The young Frenchmen who were obliged to carry out their term of compulsory military service were deployed to North Africa, whereas they had not been required to serve against the Viet Minh. "With the dispatch of the conscript contingents," Maurice Vaïsse has written, "one could believe that the nation was totally engaged in the war. In contrast to the army of Indochina, which was only an expeditionary corps formed of volunteers, eighty per cent of the army in Algeria was drawn from the conscripts, those performing their military service or retained beyond the legal limit of their service-time."[37] Thus, as one historian of the French stake in Algeria from 1945 to 1962 has put it, "within a matter of months, France found herself wholeheartedly involved in quelling the revolt by bringing about a substantial escalation of the conflict."[38] It was yet one more paradox of a perplexing war.

The second distinguishing feature of the French army's struggle in Algeria was that many senior French military commanders were directly connected to the cause of Algérie française. Some were *pieds-noirs* themselves, such as General Edmond Jouhaud of the French air force, born at Bou-Sfer near Oran, whose impassioned memoirs published in the 1970s included volumes with the anguished titles: *ô, mon pays perdu* (Oh, my lost country) and *Serons-nous enfin compris?* (Will we at last be understood?).[39]

Other leading commanders, though not *pieds-noirs*, included General Challe, whom de Gaulle appointed commander in chief in Algeria in December 1958, and General Salan, who as joint services commander in chief in Algeria had exercised a critical role in engineering de Gaulle's return to power on 13 May 1958. Salan was aloof, austere, and mysterious. A French imperialist to his bootstraps, he went by the nickname of *le chinois* or "the mandarin"—names conferred by fellow officers who admired his linguistic fluency in Laotian and Canton Chinese and puzzled over his legendary inscrutability. A romantic idealist of the empire, Salan was born in 1899 and now found himself living in the wrong era. For Salan's soul lay in the epoch of Galliéni or Lyautey. He found himself fundamentally out of place and out of time in the post-1945 world of decolonization. Congenitally incapable of imagining a France without her empire and obsessively anticommunist, Salan's sympathies were drawn as though by a magnet to the cause of French Algeria.[40]

General André Zeller, the fourth member of the military quadrumvirate that attempted the Algiers coup against de Gaulle in April 1961, was an Alsatian. By reference to his own people's experiences as a political punching bag between France and Germany, he understood perfectly how much was at stake for the Algérie française movement and their readiness to use all and any means to quell the FLN.[41] From the outset, therefore, the French war with the Algerian nationalists was, on every side, a "people's war."

The conflict in Algeria was as much a struggle over the political direction in metropolitan France as it was a contest for military superiority in France's last North African enclave. Consequently the violence escalated. The FLN staged waves of terror bombings. The French military and police riposted with increasingly indiscriminate arrests and unrestrained torture, as well as with low-flying, strafing aircraft. It was a war with the gloves off—a highly intense "low intensity" conflict of uncommon ferocity. French *nettoyage* methods and allegations concerning abuse of suspects and prisoners aroused protests of human rights violations. French actions were condemned, not only at the United Nations and by the United States, but also by an increasingly vocal opposition in Paris.[42]

Politically, the French officer–pied-noir coalition became progressively estranged from the governments in Paris. This was especially apparent between 1956 and the spring of 1958. During this phase of the war, the governments of Guy Mollet, Maurice Bourgès-Maunoury, Félix Gaillard, and Pierre Pflimlin floundered indecisively, searching for a compromise that would satisfy the FLN's minimum demands and yet preserve Algeria in the French Union. The efforts during the years from 1955 to 1960 by the ministers-resident in Algiers (or, as they were later redesignated, delegates-general)—Jacques Soustelle, Robert Lacoste, and Paul Delouvrier—to arrive at an accommodation served only to heighten the atmosphere of intrigue, suspense, and mutual suspicion.[43] Soustelle and Lacoste were to disassociate themselves from Gaullist policy once it appeared that full-blown Algerian independence lay at the end of the road down which this policy had turned by 1960–61. Soustelle indeed ultimately aligned himself with the Algérie française diehards, becoming a fellow traveler of the Organisation de l'Armée Secrète (OAS) and a member of its leadership council, the so-called Directory.[44] The question of Algeria's status was

> further complicated by the presence in the Maghreb of French army units that had seen service in Indo-China, and defeat. By 1956, when Tunisia and Morocco had obtained their independence, the bases which the French army had retained in those countries contained many officers who found the Tunisians and Moroccans could not fail to sympathize with their Muslim neighbors across the border in Algeria. Against that background, soldiers and settlers combined to destroy the regime of the Fourth Republic, in the expectation that it would be followed by a government that would safeguard the "honor" of the army and the interests of the settlers by keeping Algeria an integral part of France.[45]

De Gaulle's return in May 1958 was, of course, the stratagem of this military–pied-noir caucus.[46] The disintegration of discipline and the growing tendency to question civilian political authority among the French officer corps achieved expression by the planning of Operation Resurrection. But indiscipline and rejection of civilian policies made in Paris only emerged in their full nakedness once the Algérie française movement understood that de Gaulle himself was not a reliable buttress of the pied-noir cause. Political action became probable once it appeared that de Gaulle would not bar the surging tide of Arab nationalism and decolonization.

This unleashed not just the plotting of the generals, but it also activated conspirators of the younger generation. A new level of politicization became apparent within the French officer corps after the "Affair of the Barricades" in Algiers in January 1960, which ensued from Massu's recall to France; Massu was the last of the generals of May 1958 still serving in Algeria. "The disgrace of anyone with a distinguished combat record" would, as James H. Meisel has astutely remarked, "alarm his comrades . . . When even such a popular commander as Massu can be dismissed, for reasons unintelligible to the military mind, no officer can feel safe any more."[47] The *politiques en képi* were henceforth most active, and most dangerous to de Gaulle, among the colonels and captains. Most particularly, the activists were forming among the veterans of Indochina and of Suez, who now found themselves staring a third and even more humiliating defeat in the face.

These were unsophisticated political neophytes, men who judged the issues before them in simplistic black-and-white terms. In their twin aim of retaining French Algeria and rendering further "betrayals" from Paris impossible, they no longer recognized moral or political restraints. A disproportionately large number of them held field rank and had backgrounds in company and battalion command in the Colonial Paratroop or Foreign Legion regiments. The most prominent among them would gain notoriety as a result of their part in the struggle against de Gaulle's Algerian policy: Colonels Antoine Argoud, Hélie Denoix de Saint-Marc, Yves Godard, Roland Vaudrey, Pierre Château-Jobert, and Jean-Marie Bastien-Thiry. Most had in common an experience of frontline combat either in Indochina and Algeria (as in the cases of Saint-Marc and Godard), or at Suez and in Algeria (as in the cases of Argoud— Massu's chief of staff with the Tenth Parachute Division in 1959–60— and Château-Jobert).[48] Some died in battle for France and achieved martyrdom within the army and the uncompromising colonialist communities. Such was the case with Colonel Pierre Jeanpierre, killed leading his Legion reconnaissance group in May 1958 against FLN *fellaghas* on the Morice Line—the electrified trip-wire along the Algerian-Tunisian border—trying to deny the FLN access to eastern Algeria from their bases such as Sakiet in Tunisia.[49] For them all, the army was the French nation in historical, spiritual, and physical terms; their "personal responsibility" to the nation came before their duty to be the instruments of the government's policy, before their allegiance to the form of government, the Republic.[50]

Certain among the hard-line French field-grade officers went further still. Colonels Argoud and Bastien-Thiry eventually went underground, in Constantine and the *métropole* respectively, as leaders of sections of the OAS, the movement that repeatedly sought to assassinate de Gaulle between 1961 and 1963. Captain Pierre Sergent, whose own memoirs appeared under the unrepentant title *Je ne regrette rien* (No regrets), was a company commander of the First Régiment Étranger de Parachutistes (REP).[51] Sergent had "throughout his service in Algeria . . . gone from one disillusionment to another, ending in the conviction that the policy of the de Gaulle government amounted solely to one of desertion."[52] He, too, became a principal OAS activist after his regiment was disbanded for its support of the Salan-Challe-Jouhaud-Zeller putsch of April 1961. Captain Jean Ferrandi joined Salan as personal adjutant and spent two years furtively slipping back and forth between France, Spain, and Algeria from 1960 to 1962.[53] Lieutenants Roger Degueldre and Pierre Delhomme became the active leadership of the so-called Delta commandos, the terrorist arm of the new subversive force of the OAS that was the only weapon left to the Algérie française diehards once de Gaulle had disciplined the army.[54]

Formed in Madrid at a meeting with Salan at the end of January 1961, the OAS was linked to civilian leaders of the *pied-noir* community, such as Jean-Jacques Susini and José Ortiz.[55] Its acronym first appeared, chalked or daubed in paint on the walls of Algiers, in the following month. The fugitive military conspirators of the OAS enjoyed the backing of several very senior metropolitan politicians. Most notable among the latter was Georges Bidault, a former associate of de Gaulle's in wartime London, head of the Conseil National de la Résistance in 1943–44 and foreign minister of the Fourth Republic.[56] They also had the sympathy and media support of prominent right-wing journalists and academics. Among the latter, none was more eloquent in articulating the case for Algérie française than Raoul Girardet, a professor and writer on the sociology and ideology of the French army.[57]

By 1960–61 the Algérie française caucus was a veritable den of thieves. Pied-noir "ultras" mixed with far-right ex-Vichyites, such as the prominent lawyer and erstwhile Pétainist, Jean-Louis Tixier-Vignancourt. Meanwhile, unstable populist demagogues—many who had participated in Pierre Poujade's neo-fascist movement of small shopkeepers and artisans in 1955–56—consorted with die-hard military conservatives. From this intoxicating mixture emanated the vicious and violent challenge to de Gaulle and the specter of a decline into civil war which haunted Algeria, and indeed mainland France itself, between 1961 and 1963.[58]

Presaged by the Affair of the Barricades, the threat materialized concretely in the putsch attempt in Algiers of 21–24 April 1961. De Gaulle decisively won that power struggle, partly because of his inspired use of a television appearance in full general's uniform on the evening of 23 April during which he ridiculed the four figureheads behind the putsch as "a quartet of retired generals" and "forbade every Frenchman, and above all every French soldier, from executing a single one of their orders."[59] Backing this imaginative use of the latest medium of mass communication with radio broadcasts from his prime minister, Michel Debré, which reached into the conscripts' barracks, de Gaulle struck a fatal blow against the conspirators through what was swiftly dubbed the "guerre des transistors." The rank-and-file draftees and short-service junior officers remained impassive in the face of appeals from the coup leaders.[60] As James H. Meisel has commented, "ultimately the French army did not follow the insurgents. It . . . remained loyal, in an apathetic way. Among the officers the Few did not prevail over the Many, even though— or should we perhaps better say, because—the Many were so hopelessly divided."[61]

The second phase, the OAS terrorism that followed the collapse of the formal military coup, was arguably more dangerous. It menaced not only de Gaulle's life but also threatened to destabilize the French state and provoke a descent into anarchy, even civil war. In this phase "a demoralized, dismantled army had become an inert mass, while a minority of angry men went underground to swell the ranks of desperadoes and lost soldiers."[62] This phase of the officers' struggle against de Gaulle wore the face of murder and urban guerrilla war. It was viewed with increasing distaste by the majority of the officers who were disgusted by de Gaulle's Algerian "sell out" but nonetheless shrank from a descent into banditry and gangsterism. The phase was marked by a sordid campaign of "plastiquages," plastic explosive bombings throughout France that targeted government buildings, police stations, and military installations. It included numerous attempts to assassinate de Gaulle, culminating in the "attentats" of Pont-sur-Seine and Petit-Clamart and the subsequent arrest of the "general staff of the OAS."

But it did not require attempts on his own life to make de Gaulle understand his need to move decisively to remold the French army. He immediately grasped the importance of recapturing the allegiance and discipline of the officers. He had already taken drastic action to punish the parachutists who had sided with the generals in April 1961: de Gaulle disgraced and disbanded the prestigious First REP as well as the Fourteenth and Eighteenth Régiments Coloniaux de Parachutistes (RCP), the air force commandos, and the paratroop-commandos.[63] Nothing could have signaled more clearly to the officers that the president intended to eliminate military insubordination than this public humiliation of units that had been counted among the army's most elite and honored formations.

As the OAS reshaped itself on 19 July 1962 into a clandestine network, the members of the Conseil National de la Révolution, including Salan and his fellow conspirators, were brought to trial. Pleading for their lives at the bar of the High Military Tribunal, the defense statements made by Salan, Jouhaud, Bastien-Thiry, and Degueldre exposed how deeply their two decades of debilitating colonial retreat since 1945 had corroded and corrupted their perspectives and obsessed them with visions of communist conspiracies and governmental cravenness.[64] As Maurice Vaïsse has noted, the "determination not to give any more ground is perhaps the greatest common denominator among these men."[65] Concluding his defense, Salan haughtily affirmed that "I do not have to exonerate myself for having refused to allow communism to be established an hour away from Marseille and to have Paris put within reach of its short-range missiles."[66]

Such sentiments were echoed not only by General Zeller, during his trial on charges of sedition and armed insurrection against the state, but also earlier by the authoritative *Revue de défense nationale*. The latter published the following manifesto by an admiral in April 1959: "At grave hours, when the sovereign voice of the people can no longer express itself, the ARMY suddenly becomes aware of what it is: the People under the Flag. Then, the Army takes responsibility for the People."[67] In the aftermath of the failed coup and the penetration by the Sûreté Nationale and Direction de la Surveillance du Territoire of the OAS, the jails of Paris saw such an influx of high-ranking military prisoners that the Santé prison was described in an article by one retired general as "an annex of the Army General Staff."[68]

Bastien-Thiry, selected as an example to the other officers with OAS sympathies, was sentenced to death. De Gaulle showed no clemency toward him. At least some of the senior conspirators had to die. De Gaulle had already been opposed by members of his own cabinet in the case of General Jouhaud; de Gaulle's cabinet, led by Georges Pompidou, threatened a mass ministerial resignation if the former head of the Armée de l'Air was not reprieved. Bastien-Thiry, erstwhile rising star of the air force, polytechnician, military chief engineer, and lieutenant-colonel at the precocious age of thirty-five, was duly executed by firing squad at dawn on 12 March 1963 at Fort d'Ivry on the outskirts of Paris.[69] By showing that he would not stop at half-measures in eradicating the political malaise that had infected the officer corps, de Gaulle released the military from the narcotic of revolutionary war. For this was the desperate resort to which the frenzied ultras of the OAS had been driven by 1962–63 and which they introduced to metropolitan France in the hope of winning it over by terrorism.[70]

From the moment of his return to direct the affairs of France in 1958, de Gaulle's objective was unwavering. He kept his sights fixed firmly on the goal of embedding a new kind of civilian authority deeply into French soil. This new concept of civilian political legitimacy aimed to make a clean break from the earlier *politique des partis* of the Third and Fourth Republics.[71] In de Gaulle's eyes, the older style of politics, and the political culture that it engendered, was fatally divisive. He shared a commonly held view that the army's intervention in 1940 and in 1958 was something that the old-style politics had brought upon itself. Out went the omnipotent Chamber of Deputies, the so-called "tyranny of parties," short-lived governments, and figurehead presidency. De Gaulle, in his January 1959 constitution for a Fifth Republic, substituted a politically powerful executive presidency and a disciplined, distinctly subservient National Assembly. In the direction of the French state after 1959, the president became unmistakably the senior partner, parliament the junior. The primacy of the French presidency and its evolution into something akin to the United States model ("the President as commander in chief") was established in the 1959 constitution. It was reinforced and embedded by means of a referendum in 1962, called by de Gaulle to introduce the direct election of the president by universal suffrage.

The new regime's primary task was to reestablish France as a major power, and especially to consolidate a position for France as a leader of the new European Economic Community (EEC). The essential precondition for accomplishing this objective, judged de Gaulle, was the reassertion of political control over the military. To bring this about, de Gaulle understood that he needed to purge from the French army's officer corps those dissidents who opposed his negotiations with the Algerian nationalists of the FLN in favor of Algerian independence. De Gaulle perceived that nothing would be achieved until he had cleansed the army high command and field officer corps of its hankerings—its "fantasmes" to use de Gaulle's word—for a quasi-institutionalized role as arbiter in French politics.

De Gaulle's strategy to bring about the recovery of the French armed forces in the 1960s commenced with his determination to erect a civilian regime whose authority would be incontrovertible in the eyes of not only the people but also the military. It was to be a civilian authority that would inspire respect as well as obedience from the top to the bottom of the officer corps. The Fifth Republic—de Gaulle's republic when all was said and done—was to represent a renaissance of the republican concept of the state. It was a renewal, surmounting at last the Bonapartist corruption in Caesarism of the original Jacobin republican authority of 1792–94.[72] Indeed de Gaulle quite deliberately and judiciously entitled the volume of his memoirs dealing with the early 1960s "Le Renouveau."[73] "It was," as one scholar reflects, "the tragic mission of de Gaulle to be both the restorer and the wrecker of French power."[74]

But the Fifth Republic's brand of republicanism was more than merely a renewal, more than a phoenix rising from the ashes left by the incineration of the Troisième and the Quatrième. Whereas the Fourth Republic had been fashioned quite consciously as the offspring and imitator of the Third, de Gaulle placed the emphasis in his constitutional reconstruction on change rather than on continuity. After 1958, therefore, there was no longer that "diminution of the state" about which so many senior officers, among them de Gaulle, had complained during the 1930s, 1940s, and 1950s. Capped by the ordinance of 7 January 1959, the role of "tutor of the nation" was formally ascribed to the civil power in the person of the president of the Republic, initially de Gaulle himself. The military were, at least metaphorically, sent back to their barracks.[75] "The purges following the barricades revolt of 1960 and the generals' revolt of 1961," it has been argued, "hit the French army with the impact of a major social earthquake. The rotation of commanding officers accelerated at a dizzying speed; the massive turnover at lower levels, the disbandment of old, glorious regiments—it all amounted to a real revolution."[76]

In 1958–59 de Gaulle proved himself adept not only in staging his putsch or coup via Operation Resurrection, whereby he claimed for himself the "mantle of national legitimacy which I have incarnated for twenty years," a legitimacy ridiculed by his opponents—and his victims, including Bastien-Thiry—but a legitimacy that effectively linked the French nation to its regime for the first time since the beginning of the Terror and the Vendée uprising in 1792.[77] Disciplinary actions and investigations after the failed Algiers putsch of April 1961 had affected 200 officers by June that year. Of those officers, 110 were subsequently indicted. On 1 July 1961 the Eleventh Airborne Division's transfer from Algeria to a new permanent base on the mainland, at Toulouse, was announced. Various generals, even those such as General Fernand Gambiez who had been temporarily arrested by the April 1961 plotters, were reassigned by de Gaulle or retired.[78]

New men were rapidly promoted to guide the French army back to discipline, duty, and a new Euro-centric mission. De Gaulle systematically gave leadership assignments to generals with reputations as reforming military intellectuals. The new commanders were strategic thinkers such as General André Beaufre or scientifically qualified technocrats such as General Pierre Gallois of the air force ("godfather" of France's nuclear force). Another, rewarded for his loyalty during the Algiers putsch, was General Charles Ailleret, whom de Gaulle appointed as joint-services commander in Algeria in the final months of French control, from April to June 1961, and who was subsequently, as chief of the staff of the armed forces in the later 1960s, the architect of France's notorious "all-round nuclear targeting" doctrine of *défense tous-azimuths*.[79]

The advent of French nuclear weapons capability offered the clearest opportunity for de Gaulle to wrench the French officer corps clean away from its contaminated, corrupted past of self-absorbed political interference. De Gaulle successfully constructed a "consensus" on the overriding importance to France of independent nuclear forces (the *force de frappe* or, as it was redesignated in the less-provocative language of the 1980s, the *force de dissuasion*).[80]

This force was first designed as a crude strategic nuclear retaliatory capability of last resort. It was, at the outset, an exclusively Gaullist conception—a new means at the disposal of French security that would be indelibly linked to the vision of French "grandeur" personified by de Gaulle. Like France's permanent membership on the United Nations' Security Council, as well as the French zone of military occupation in both Berlin and southwest Germany that de Gaulle had wrested from the United States, the Soviet Union, and Britain in 1945, the force de frappe constructed in the 1960s was a totem. It symbolized the new political virility of the Fifth Republic. It spoke of the risen, resurrected France—a Gaullist France.[81]

But not the least of de Gaulle's achievements in the realm of national security policy by the time of his resignation from the Elysée in April 1969 was his securing of a cross-bench political consensus around this singularly French version of a nuclear deterrent. Under the subsequent presidencies of Pompidou and Giscard d'Estaing (1969–81), the nuclear capability in the French armed services enjoyed overwhelming public approbation. Not only this, it had the seal of party political approval across the spectrum from the reorganized Gaullists of Jacques Chirac's Rassemblement pour la République to Georges Marchais' Parti Communiste Français. During the presidential election campaigns of 1974, 1981, and 1988, not one of France's major political parties still campaigned against French possession, modernization, and retention of a *force nucléaire*.[82]

De Gaulle set the seal on this reconstructed consensus or solidarity between nation, regime, and armed forces. Parties and populace alike rejected the rebellious officers of the die-hard Algérie française cabals, so that de Gaulle's purges of 1962–63 were once-and-for-all operations. Absolutely central to this reassertion of military discipline and civilian political primacy was de Gaulle's redefinition of the French military's function or role. The buildup of the French nuclear force was the single most important element in this recalibration of the French military's mission.[83]

De Gaulle, in short, was not simply another "man on horseback." He did more than reincarnate the Caesarist or Bonapartist tradition that has run through French political fashion since 1789 in counterpoint to republicanism. De Gaulle was too shrewd to envisage his regime as a modern form of presidential dictatorship. "Can anyone think that I am about to embark on a career as a military dictator at almost seventy years of age?" he asked in a speech to reassure and rally republican waverers during the referendum of September 1958 on the Fifth Republic's constitution. De Gaulle, notwithstanding the triumphalist glorification of de Gaulle's person, so sedulously cultivated from 1958 to 1969 by the Gaullist propaganda machine, was a democratic elitist, a French Washington, not a Corsican demagogue.[84] He knew full well that the French had already endured one bitter and divisive modern experience of government by military politicians through the self-styled "National Revolution" of Pétain's Vichy regime.

What de Gaulle carried out was nothing less than the reprofessionalization—and thereby the depoliticization—of the French officer corps. By providing them with a new sense of mission, de Gaulle broke the military's need to control the direction of the state. De Gaulle restored the generals' trust in the "reliability," the patriotism, the efficacy of a democratically elected and accountable presidential-parliamentary order. In doing so he obviated the need for the old military propensity to differentiate between nation and regime. Removed from a role as arbiter, broker, and king-maker in politics, the officer corps gained a revitalized sense of self-worth.[85] For they were entrusted with executing de Gaulle's ambitious program to deploy the military atom in the service of the security of France.

The theoretical feasibility of the Gaullist nuclear force received timely and crucial impetus toward technical realization in February 1960. In that month, at Reggane in the French Sahara, the first French atomic device was successfully exploded. Shifted from a theoretical proposition of nuclear physics and experimentation to that of a usable weapons system, the French nuclear force became capable of fulfilling an essential need—the need to redefine the vocation of the French officer corps. With what may, with hindsight, seem to have been a providential coincidence in time, French nuclear pretensions moved a huge step closer to reality just when tensions over disengagement from Algeria were mounting to their convulsive climaxes on the barricades in Algiers in January 1960 and in the abortive putsch of April 1961. Doubt-ridden, weary, and angry, assailed by a sense of failure and betrayal by metropolitan France, the French officer corps, especially that of the army, was near the end of its tether by 1960.

Suddenly a new agenda offered itself. The harnessing and application of the atom provided a route back into a profession from which a generation of disenchanted and embittered career officers could draw self-respect. A highly prestigious career, a veritable calling or vocation, suddenly beckoned in the bright, shiny new world of nuclear defense. This new world of the Gaullist and post-Gaullist military would cease to have any connection with the former, discredited world of the colonial casernes. It would be singularly free of political ambitions. In place of colonial grousers and *politiques en képi*, the reformed French armed forces came to be composed of uniformed technocrats. The land was bright again. As Patrice Buffotot has put it: "The nuclear arm provoked a veritable transformation in the French army, which passed from an army of the 'colonial' type in which the foot soldier was the principal element to an army of technicians at the service of sophisticated weapons-systems."[86]

De Gaulle had striking success in a comparatively short time. Between 1962 and 1966 he largely accomplished his goal of refocusing the concerns of French officers on their purely military function, their professional mission. In little over three years the insidious predisposition of the more senior French officers to make their own political judgments had been exorcised. In a fundamental way, the French army had been resubordinated to the civilian authorities. Eloquent testimony to this reversal of positions was the army's dethronement from a previously privileged association with the state power at the highest level. The termination of the military tendency toward political interference was most clearly gauged by contrasting this with the retention of influence of other quasi-corporatist special interest groups. Comparatively speaking the army lost ground relative to the influence retained by other national collectivities, such as the agriculturalists, the trade unions, and patronal organizations, the Confédération National du Patronat Française, and the Association des Petites et Moyennes Entreprises.

De Gaulle was extraordinarily adept not only in purging but also in refocusing the French armed forces around a clear, unambiguous, and attainable mission of identifiably "national" character. This set the seal on the reconstructed solidarity of regime and nation. The nuclear force was originally meant to apply only to direct threats against France. It thus implied that France would cut its military ties to NATO's integrated command structures. This started with de Gaulle's withdrawal of the French Mediterranean fleet from the NATO Mediterranean control in 1959 and led to the expulsion of the NATO secretariat and the Supreme Allied Commander Europe (SACEUR) headquarters from Paris to Brussels and Casteau in Belgium in April 1966.[87] The French armed forces were given a clear and uncontroversial role: the "sanctuarization" of metropolitan France against nuclear attack and the maintenance of France's best-equipped army corps in Baden to assist, in last resort, in a land defense of the eastern approaches to the Rhine against Warsaw Pact aggression on NATO's central front.[88]

Conclusion

The rapid recovery of the French army after its demoralizing and discipline-shattering debacles in Indochina and Algeria remains among the pivotal achievements of de Gaulle's decidedly remarkable career. It was possible in 1962 for one scholar of the malaise that had then torn asunder the French military professionals to qualify Jean-Raymond Tournoux's saying that the French officer corps "is a class" with the suggestion that, though this might be the case, "it has not yet developed much class-consciousness."[89] This consciousness was, however,

102 Martin S. Alexander and Philip C. F. Bankwitz

precisely what de Gaulle, his ministers for the armed forces, and their successors brought into being.

The French officer corps was permitted to redefine itself, to emerge from the time warp of colonial war in the late 1950s, shedding its old, rigid carapace of fixations concerning its nature, history, and destiny that had developed during the long nightmare of internal social and political dissension and international peril of the previous hundred years.[90] Within a mere ten years—and arguably in less time than that—it had been transformed into the master of the complex, technocratic, and scientific milieu of the late-twentieth-century industrialized defense policy, strategic doctrine, and force planning. In the extraordinarily apt summation offered by Michel Martin, the French officers in effect transmogrified themselves from "warriors into managers."[91]

This process permitted them to reinvent their definitions of professionalism and rediscover their own code of duty, obedience, and honor. De Gaulle assigned the French armed services, and most crucially their career officers, a clear mission: the army was to play its part in the wider panoply of French armed forces as providers of security against the Soviet Union and the Warsaw Pact, through threatening an early escalation of conflict in Europe across the nuclear threshold. In the new concept of war, the alluring "horizontal" part, involving politics and ideology, to which the officer corps had become addicted during the colonial conflicts, now finds its proper place alongside the older "vertical" one of nation-state struggles between organized military units. Revolutionary war becomes the complement of, not the antithesis to, nuclear war.[92] Once their role had been thus reconceptualized in austerely functional terms—and once de Gaulle and his minister for the armed forces, Pierre Messmer, had conveyed that the civil power backed the uniform professionals in what they were being asked to do—the French military cadres regained a raison d'être. Politically, as one scholar writing in the late 1960s put it, the French officer corps "returned to silence."[93] Recovery from defeat in the French case could be as rapid as it was because the Fifth Republic made the armed forces once more worthy of an officer's dedication of a professional lifetime in their service.

6

. . . .

Defeats and Recoveries in the
Arab-Israeli Wars of 1967 and 1973

Amos Perlmutter

Historians and military analysts frequently disagree about the Arab-Israeli wars, even to the extent of arguing over their actual number. Numerous questions are debated: Did the wars begin in 1920, 1936, or 1947? Is the retaliation warfare between 1949–1953 a continuation of the 1947–1949 War of Liberation? Is the Sinai War of October 1956 the last war of the first round, or the first round of the next Arab-Israeli war?

I prefer to focus on a later period: on the nature and character of the consequences of the 1967 and 1973 wars. My theory is that the 1967 and 1973 wars are a continuation of the same war, linked by the 1968–1970 War of Attrition. What I find particularly striking about this period is the amazing amount of continuity of personnel. Most of the key Israeli and Egyptian political and military *dramatis personae* took part in all three wars.[1]

A great number of senior officers in the Israeli Defense Forces' (IDF) high command participated in all three wars. Moshe Dayan presided over all three wars as defense minister. Yitzhak Rabin was the chief of staff in 1967. Chaim Bar-Lev was deputy chief of staff in 1967, chief of staff from 1968 to 1972, and David Elazar's (Dado's) deputy in 1973 as well as chief of staff of the Southern Command. Elazar was a senior officer in 1967 and 1968–1970 before becoming chief of staff in 1973. General Ariel Sharon distinguished himself in all three wars as did Generals Israel Tal (deputy chief of staff in 1973) and Aharon Yariv (chief of intelligence in 1967 and Israel's chief military negotiator with the Egyptians in Kilometer 101 after the 1973 War). Sharon, Tal, and Bar-Lev all served at some time as commanders in chief of the Southern Command. The number of brigadier generals, colonels, and lower-ranking officers whose careers were fueled by participating in the three wars is considerable. Since 1973, there has been no chief of staff who did not participate in at least one of the wars, sometimes two or three. Almost all of the post-1973 generals were graduates of the three wars, serving as junior or midlevel officers.

The 1967 War was, in military as well as political terms, a watershed event. What distinguished the 1967 War from past Arab-Israeli wars was that it was fought on the basis of a new strategy by a highly professional IDF, which had developed into a modern, cohesive armed force equipped with modern weapons, armor, and an air force. The lessons of 1956 had clarified the importance of a large-scale army based on newly formed divisions, with an armored corps and an independent, aggressive air force. It was the predominance and efficiency of the armored corps and the air force that allowed the Israelis to annihilate the Arab air force, most of it on the ground, and to blitz into Suez, the Jordan, and the Golan Heights.

Over the years, the infantry army of 1948 and 1956, along with the paratrooper force that had so distinguished itself in the War of Retaliation of 1949–1956, had been reduced in size, but the paratrooper élan and concept of leadership had been transferred to the growing armored and air forces. In effect, its members spiritually amounted to paratroopers in tanks and jets. The aggressive and winning spirit of 1948 and 1956 had become mechanized to the point that by 1967, Israel had perhaps the world's finest armored and air forces in terms of fighting spirit and capability. This spirit would make itself felt in the strategy of Israel's military chiefs and commanders.

If the IDF strategy was cohesive, coherent, and successful, these qualities were mirrored in the continuity and longevity of Israel's political elite. The three wars were conducted under the political leadership of Mapai-Labor, which had governed Israel and the Yishuv in Palestine since 1933. The two prime ministers of the period, Levi Eshkol and Golda Meir, were the heirs of David Ben Gurion, the founding father of the state and the IDF, whose historical influence could still make itself felt. The senior ministers of the time had wielded power in one form or another since 1948 and some of their careers went as far back as the diplomatic 1930s and the pioneer 1920s. This was a first generation of founding fathers (and mother, in the case of Meir), whose authority was wide, whose legitimacy was unquestioned, and whose power was considerable.

In addition to the two prime ministers, some of the key individuals in Israeli politics were Israel Galilee, Moshe Dayan, Yigal Allon, and Pinchas Sapir. Galilee, Dayan, and Allon had been outstanding Haganah, Palmach, and IDF leaders. Galilee, in fact, was a founding father of the Haganah-Palmach and had been Ben Gurion's deputy in 1947–1948. Dayan had been chief of staff of the IDF from 1953 to 1956, in addition to being Ben Gurion's protégé. At home and abroad, Dayan, dashing, handsome, and charismatic, had become the symbol of the Israeli officer corps and its fighting spirit. He entered the cabinet during the "waiting" period in May and early June and became defense minister just prior to the start of the 1967 War. Allon had been a distinguished commander of the Palmach, which had defeated the Egyptians in the War for Independence. He had been a senior minister and was Dayan's chief rival for Labor party leadership. Galilee, a minister without portfolio, was considered the éminence grise of Israel's security and served as Eshkol's and Meir's chief defense advisor and speech writer on security matters.

Pinchas Sapir, the treasury minister, was never really a part of the Meir-Dayan-Galilee troika of 1969–1974 (Eshkol had died in 1969), but as the prime minister's financial advisor, he wielded considerable budgetary influence in defense matters. In addition, there was Rabin, the Israeli ambassador to the United States in 1968, who had considerable diplomatic power even though he had little voice in strategy and cabinet-level security considerations. The political troika and those around them established political and diplomatic policies and imposed their own political strategies on the IDF, an arrangement not always welcomed by the IDF command.

To understand the startling success of the IDF in 1967, the equally startling fiasco of 1973, and the IDF debate over strategy in the War of Attrition, we must analyze the political, intellectual, and security doctrines of the ruling troika, as well as the relationship between these doctrines and the debate over security after 1967 within the IDF high command, a debate that would not be resolved until 1973. Then we will need to examine military and tactical strategy.

To explain the aspirations and weaknesses of the political elite, we must look for the philosophical and practical roots, the Weltanschauung, of Israel's inner circles, the troika, and others often referred to as "Golda's Kitchen."[2] Anchored in the tradition of the socialist-Zionist, Mapai, and Pioneer concept of the elite, that the rule of the few should guide the nation, Israeli national security policies were conceived and implemented by a small, informal, and familiar group. They were the "dedicated" members of Labor, of Mapai, or of the Histadrut-Hityasvut movements that had established the economic, political, and security institutions of the nation from the early days of immigration to Zion onward. Now they guided the affairs of state, especially those concerned with security. They comprised a small, coherent, and tightly knit circle known as the *Bithoniim* (the security old hands) who specialized in Yishuv and state security affairs and were also the coalition leaders.

Their political culture was agrarian and socialist and was oriented toward land acquisition. Dayan and Allon, for example, grew up in cooperative agricultural settlements. Allon was a kibbutz member and belonged to the territorialist and hawkish Achdut Haavoda (United Labor) party, which finally merged into Labor in 1968. For them, territory was space, the essence of security. Territory also allowed them to deal with the Arabs from a position of strength, a lesson they had learned from the early Pioneer settlers who combined agricultural labor with military security. As a group, they considered themselves "pragmatic hawks," especially their leader, the single-minded and implacable Golda Meir whose position after 1967 was crystal clear—not an inch of territory until the Arabs were willing to negotiate.

This position produced the Labor party's "peace for territory" formula. Thus, the 1973 strategic doctrine of the inner circle evolved from the results of 1967. The Arabs, correctly, interpreted "peace for territory" as meaning no serious territorial concessions. In Egypt, the troika's security doctrine was seen as a policy of annexation under the guise of verbal compromise.

Between 1969 and 1974, the political troika gave only general guidance to the IDF concerning its military doctrine, force structure, deployment or levels of strength, and weapons acquisitions. The IDF's military doctrine was intellectually and doctrinally subordinated to the troika's political conceptions and territorial aspirations. The military professionals clearly understood the concept of civilian domination and tried to ensure the security of the nation using the rigid territorial political doctrines with which they had no serious intellectual quarrel in any case. In some ways, the military were clones of the political elite, who in some cases, came from the ranks of the military.

In spite of the opportunities presented by victory in the Six-Day War, Israel's political leaders remained locked into pre-1967 territorial security concerns and apprehensions. In other words, the period between

1967–1973 was a history of lost political opportunities and the military's failure to challenge the strategic wisdom of their civilian superiors. Let us look at some of the opportunities created as a result of the swift Israeli victory in 1967.

The original boundaries of the state of Israel were a product of the United Nations' partition and the War of Independence, that is, a combination of international legitimacy and the results of the use of force. Even though the War of Independence somewhat improved Israel's original boundaries, strategically there was not much territorial breathing room. The 1967 War radically altered Israel's frontiers and boundaries: to the east was a natural waterway, the Jordan River; to the south, the Suez Canal, and to the north, the Golan Heights. Defense Minister Dayan declared on 11 June 1967 that Israel's borders were now "ideal." There was enough room in which to maneuver politically, as the 1979 Israeli-Egyptian peace treaty demonstrated. In Israel's mind, territory (although not all territory) could be exchanged for real peace, meaning a peace treaty.

After the 1967 War, no Arab leader or combination of Arab leaders could seriously contemplate an easy or complete destruction of Israel by military means, even though ideologically and rhetorically all Arab states remained dedicated to the proposition. The purely military option of annihilating Israel no longer seemed realistic.

Since 1949 the IDF's military doctrine had been based on deterrence and bringing the war into enemy lines, as was effectively practiced in 1956. After 1967 this was no longer necessary. In point of fact, the strategic-defensive replaced the offensive-deterrence strategy. The psychology of encirclement as experienced by a garrison state was no longer a reality. The new territorial conditions eased the psychological apprehension over Arab aspirations and the threat of annihilation that had been the norm in the 1950s. Real deterrence had come of age: "Working on the psychology of the enemy so that he will not decide to attack."[3] In a nutshell, this had become the Israeli political-military doctrine in the aftermath of 1967.

This concept—shared alike by the troika, the government, the coalition, the opposition, the military, the nation, and society—was rooted in the idea that there was no pressing need for early and extensive territorial concessions. Thus did "not an inch of territory" prevail. The idea was to implant in the Arab mind the idea that territorial concessions would be made only from a position of Israeli military and political strength. To the

Israelis, territorial concessions meant political weakness. An unrealistic feeling of invulnerability emerged in the heady wake of victory. Generals were lionized, and a rush of generals into politics became fashionable for a time. Labor, under Golda Meir, made no serious effort to indicate that Israel was amenable to compromise, diplomacy, or negotiation. This elite clearly represented Israeli public opinion and consensus: peace for territory.

The military-strategic doctrines after 1967 were derived from the doctrines of the political elite and a national consensus on territorial determination. Thus, the transformation from a purely offensive strategy to a mixed, offensive-defensive policy after 1967 characterized the political consequences of 1967 and the failure of the political leadership to turn a brilliant military victory into an enduring peace. The Arabs were not willing to be accommodating or cooperative either and would not do so until Egypt was ready for political compromises in 1977. The new geopolitical and strategic conditions had considerable effect on the IDF, which was indoctrinated to "defend what exists." The definition of "what exists" was amended, however, to include the "fruits" of the 1967 war, the ideal borders.

These borders are best represented by the Bar-Lev line.[4] The Bar-Lev line was a product of the Egyptian War of Attrition and soon enough became the IDF strategic defensive line with fortifications (*Maozim* and *Taozim*) that included armored divisions, artillery, and a road system around the canal and in southern Sinai to protect the forces on the canal.

As Bar-Siman Tov clearly argues, "The construction of the Bar-Lev line was indicative of a shift in Israel's security doctrine. Even more so it symbolized its desire to perpetuate the territorial, *political and military status quo."*[5] The line was as much a symbol of Egypt's determination to challenge Israel as it was of Israel's status quo. Egypt's major war efforts between 1968 and 1973 amounted to attempts to breach the line and to wreak havoc on the IDF forces on the line. Bar-Siman Tov summarizes the factors that contributed to the construction of the Bar-Lev line, or more accurately, the Bar-Lev IDF strategy: (1) the presence of IDF forces on the eastern bank of the Suez Canal; (2) the consolidation of the territorial status quo in the Sinai; and (3) the need to defend the troops positioned on the canal.[6]

An Israeli cabinet decision dated 19 June 1967, only days after the great victory, confirmed both the political and military strategy—"Israel will retain the territories currently under its control until a peace treaty is signed with Egypt." Yet the debate over the Bar-Lev line that took

place after its construction was not so much over the territorial imperative as it was over the most efficient way to implement this imperative.
The debate raged fiercely among the generals of the 1967 War and the
War of Attrition, but it was basically tactical in nature. The debate represented the thinking and tactics of the senior officers who participated in
all three wars. The 1968–1972 War of Attrition contributed mightily to
the thinking of these senior officers up and down the ranks.[7]

The Bar-Lev strategy was conceived and established on the basis of
the experiences of the War of Attrition. Briefly, the War of Attrition ran
from 1968 to 1970 and constituted the next phase of Israeli-Egyptian
warfare which, technically, began on 11 June 1967 but gained momentum and purpose only after 1968. It was a static war along the canal that
involved an Egyptian attempt to wear down the Israeli forces strung out
along the newly built Bar-Lev line. What was important about the war
was the massive Soviet intervention in the training, manning, and equipping of Nasser's battered and defeated army. The War of Attrition was
Nasser's first phase in a new military-psychological campaign against the
IDF forces stationed on the canal. The *renewed* construction and fortification of the line, in fact, its institutionalization, was the IDF's response to
Nasser's strategy.

To explain the War of Attrition and the Egyptian strategy that would
eventually lead to the Yom Kippur War of 1973 as well as the Israeli political and military response to it, one must deal first with the key Egyptian
military and political *dramatis personae*.

Without a doubt, the two most important Egyptian figures in the
period between 1967–1973 were the two military dictators, Gamal Abdul
Nasser and Anwar al-Sadat. No Egyptian group of generals and politicians corresponded to the Israeli political troika, group of ministers,
and IDF high command and senior officers. Egypt was, and remains, a
praetorian dictatorship.

Nasser, who ruled from 1953 until his death in 1970, was unquestion-
ably the most prominent Pan-Arab and Egyptian nationalist of the time.[8]
Nasser's ideological commitments were directed toward unifying and
establishing Egyptian hegemony over the emergent postwar Arab world.
Nasser was a supremely confident and able political strategist, but his
military experience was limited to his post as a lowly major in the Egyptian expeditionary force that invaded Palestine in 1948. Nevertheless, he
was a remarkably skilled plotter.

Vatikiotis is certainly correct when he says that "it is futile to try to establish whether or not Nasser believed in a political doctrine of Arab nationalism."[9] What is significant is that he became the first Egyptian ruler to involve himself in Pan-Arab politics. The instrumentality of Arab politics drew Nasser into the Palestinian question and thus eventually into a confrontation with Israel.[10] He turned Egypt into a flagship for Pan-Arabism. In his search for an Arab role, he became involved in the Palestinian-Israeli war. From an ambivalent attitude toward Israel before 1953, he became fully involved in Israeli-Egyptian retaliatory warfare, which climaxed in the 1956 Sinai war. Even so, he was not ready for a direct attack on Israel, neither in 1956 nor in 1967.[11] In 1967, his brink-manship triggered Israel's preemptive response, igniting the war. He was "perhaps misled into such confident complacency because he was en-couraged in his brinkmanship by the seeming success of his diplomatic initiatives that month (May)."[12] Nasser assumed that the Soviet Union would offset U.S. support of Israel, or at least afford a restraining influ-ence, and that his armed forces, reorganized and equipped by the Soviets after 1964, could meet the test. In all of this, he miscalculated disastrously.

The crushing 1967 defeat was the beginning of the end of Egyp-tian leadership of Pan-Arabism. It might have deterred a lesser man but, far from being despondent or losing hope, Nasser remained defiant. At the Arab Summit Conference in Khartoum in November 1967, Nasser issued his ringing three no's: no peace, no negotiation, no recognition.

This was the beginning of Nasser's War of Attrition, which Sadat would continue in 1970–1973.[13] The goals of Egypt's political, military, and psychological strategy of 1973 were sown at the end of 1968 and were tentatively exercised during the War of Attrition along the Bar-Lev line. The central aim of the War of Attrition was to violate, to disrupt, and to end the Israeli doctrine of the status quo. The War of Attrition officially came to an end through American mediation on 8 August 1970 and re-sulted in a situation that was neither peace nor war, a situation intolerable for Egypt but comfortable for Israel. "The world already lost confidence in us," Sadat wrote in 1974, "and we began losing it ourselves . . . The state of neither war nor peace," he continued, "was prolonged and we have found ourselves in conditions similar to 1948 when our acquiescence meant the stabilization of Israel's frontiers. Israel and the world became convinced the Arabs will not fight and thus there is no reason for Israel to withdraw from its new conquest."[14] The Egyptian decision to go to war was made sometime in 1971, after a long period of trial and tribulation between the newly installed Sadat and a very small group of political and military advisors.

Sadat's small group of military advisors played a crucial role in translating Sadat's political doctrine into the details of military planning. The group was led by General Ahmed Sidki Ismail, the minister of war, and included General Abdal-Ghani al-Gamasi, director of operations, General Muhammed Ali Fahmi, chief of air defense, and General Saad al-Din al-Shazli, the chief of staff. All of them were highly professional soldiers. Together, working meticulously and steadfastly over a period of three years, they planned the operation of crossing the canal in every detail.

Sadat was not in the least like Nasser. He lacked charisma and was hardly known to the outside world. He had been a senior member of the Free Officers Club which led the coup of 23 July 1952, but he never held any important political position in Nasser's many and varied cabinets. Nasser once referred to him as "Colonel who?" Sadat, underestimated by friend and foe alike, proved to be a crafty infighter and survivor. In 1970, to the surprise of many, he took it upon himself to assume Nasser's mantle after the dictator's unexpected death. Hardly anyone expected him to last in power, and certainly no one foresaw that he would become a major international figure, equal to, or perhaps even surpassing, Nasser in achievement. Sadat continued Nasser's effort to build an effective army that would be ready to "eradicate the traces of aggression." But how? It would be effected through a political-psychological-military doctrine that would puncture Israel's overconfident security posture and doctrine. "We must confront Israel's security doctrine in order to destroy it completely," he said.

The military's mission was relatively simple. "The mission of the Army," wrote Sadat, "was to destroy as much as possible of Israel's armed forces, since Israel rightly or wrongly was convinced its armed strength will deter the Arabs. To destabilize this doctrine only a massive and destructive military operation is necessary."[15] Sadat and his minister of war, General Ahmed Sidki Ismail, pondered the problems of the Israeli military doctrine with their senior officers, including Generals Abdal-Ghani al-Gamasi, Saad al-Din al-Shazli, Hosni Mubarak, Hasanain Heikal, editor of *Al Ahrem*, and General Ahmed Abu Ghuzalah (later to be Egyptian defense minister and Mubarak's number two man).

In the Egyptians' view, the Israeli military doctrine was built around the following elements: (1) decisive military and technological superiority; (2) splitting the Arab military potential by moving from one front to another; (3) the quick transfer of war to the enemy's territory; (4) a quick, blitz-like war; and (5) maximization of the enemy's casualties. To counter this strategy, the Egyptians came up with the following operational plan:

(1) make Israel fight on two fronts; (2) inflict heavy human and material damage on the IDF; (3) create conditions for a prolonged war that would fully mobilize Israel and quickly become economically and socially intolerable; and (4) mobilize full Arab solidarity in order to use the most effective and economic means at their disposal.[16]

The Egyptian war minister General Ahmed Ismail and his highly professional staff developed a detailed attack plan. The very nature of the planning and planners was indicative of the changes that had occurred in the Egyptian armed forces. Before 1967, Nasser's military chief was Field Marshall Abd al-Hakim Amer, a veteran of the Free Officer Club who ran the high command and senior staff on the basis of political patronage, where political affiliation took precedence over professional ability. Sadat's army was representative of Nasser's purges after the 1967 War. Further reorganization after 1970 called for greater professionalism and opened the ranks of the officer corps to college and university-educated officers, to the middle peasantry, and to members of the working class. The army was no longer privileged, as it was before 1952, or heavily politicized, as it was under Nasser.

But the biggest influence and impact on the Egyptian armed forces came from the Soviets. Soviet professional advisors trained the Egyptians and instilled a high degree of professionalism and discipline, both of which were previously lacking. Egypt's army underwent exercise after exercise, until each soldier knew his duties by heart. The Soviet military doctrine of the massive-collection drill was successfully adapted to the Egyptian army.

The Egyptian plan was designed to counter the strengths of the IDF and to take advantage of its inherent vulnerabilities. If the IDF infantry were to take a back seat to its armored corps and air force, Egypt would throw against it an army that was massively tilted toward the infantry. If Israel were to excel in tank warfare, Sadat and Ismail would employ their newly devised form of antitank warfare, using the 50,000 Soviet-made RPG's that had arrived in Egypt just before the start of the war. If Israel dominated the skies, Egypt, along with Syria, would try to counter that advantage with the use of Soviet-manned antiaircraft missile batteries. Just before October, the Soviets moved their missile batteries to within thirty kilometers of the canal, in spite of protests by the United States and Israel. The aim of the Egyptian master plan was clear: "To cross the canal of shame."

The considerations and principles of the Egyptian master plan closely followed its political-psychological doctrine: to inflict heavy damage to

Israel's most prized asset—manpower. Israel could not tolerate either massive casualties or a prolonged war.

But the key element of the Egyptian plan was the element of strategic surprise—that is, to break Israel's Bar-Lev line and cross the canal while inflicting heavy damage on the relatively small Israeli forces along the Bar-Lev line, and carry on a two-front war with Syria helping to split the IDF in the north. The first forty-eight hours were meant to consolidate the crossing of the canal. The master plan did not include an Egyptian advance into Sinai or Israel proper.

On the operational level, a surprise attack on two fronts and immediate seizure of territory had to be accomplished in order to maximize the effectiveness of the massive Egyptian army. A large-scale air defense would restrict the Israeli air force and neutralize it while establishing a defensive bridgehead against the expected Israeli counter-offensive, thereby restraining the IDF's offensive capability and inflicting heavy casualties.

To achieve a complete surprise, the Egyptian high command needed to counter the brunt of Israel's armored divisions before the IDF fully mobilized (a matter of only twenty-four to forty-eight hours). To meet this challenge, according to Muhamed Hasanain Heikal, Nasser's closest advisor, the Egyptian command equipped the first assault infantry force with "the best portable anti-aircraft and anti-tank missiles in the world, the Soviet Strellas and Molutkas (known elsewhere as SAM-7s and Saggars). The solution was to equip the first 8,000 shock troops and infantry divisions which went in immediately after them, with these missiles on a scale far in excess of anything previously contemplated." [17] This is an exaggeration on Heikal's part, but the idea and nature of the assault were certainly part of Sadat's plan.

On the whole, Operation Badr was successful. At 1405 hours on Saturday, 6 October 1973, 4,000 guns, rocket launchers, and mortars began firing along the Egyptian front, with an additional 1,500 guns erupting along the Syrian front. A rolling artillery barrage accompanied some 8,000 troops who were crossing the canal using 1,000 rubber boats. Elements of the Second Army captured the first fortresses of the Bar-Lev line only an hour later at about 1500 hours. Others fell soon thereafter. By 1930 hours, the first formations of two Egyptian armies were established on the eastern bank of the canal. About 80,000 men in twelve waves had penetrated Sinai and dug in next to the canal. Sadat's basic goal had been accomplished: his forces had crossed "the canal of shame."

The psychological damage to the IDF and Israel was enormous. To this day, the war is called the *Mehdal,* or misdeed; Zeev Schiff, the defense analyst, entitled his book *The October Earthquake.*[18] The Arab "wall of fear" had been dissolved. The Egyptians and Syrians had achieved strategic surprise accompanied by psychological shock by splitting the IDF into two fronts on the same operational level, giving them a significant political advantage. In fact, the war ended with an Israeli counteroffensive that destroyed some 800 Egyptian tanks and that saw the Israelis cross the central western canal and encircle the Egyptian Third Army on the eastern side of the canal. The Syrians, who had successfully penetrated the Golan Heights, were repulsed by an Israeli advance into Syrian territory that stopped just short of Damascus. Yet despite being thirty kilometers from Damascus and a hundred kilometers from Cairo, the Israeli counteroffensive gained no political or strategic advantage. The invasion of Israel, of course, had never been part of the Sadat-Assad plan.[19]

Militarily, the most important results of the Yom Kippur War were determined on the first day when Egypt crossed the canal and the Syrians penetrated the Golan Heights. On the third day, 8 October, when the Egyptians repulsed an Israeli counterattack, they completed and institutionalized the results of the first day. IDF efforts to minimize and limit the Egyptian and Syrian advances were not successful. The Israeli line which held for seven years along the Suez-Sinai and the Golan was not merely penetrated, it was demolished. This was the true nature of the Egyptian-Syrian strategic success, and ensuing events could not alter it.

In the Yom Kippur War, the IDF was led into a war it had not decided to fight, a war in which it could not prevent the realization of Arab advantages. One had to ask: why? To explain the Israeli military's misperceptions and strategic inconclusiveness, we must return to an analysis of the war among the generals, the debate over the Bar-Lev line, before, during, and after the war.

In November and December 1968, the debates raging in general headquarters concerned the options facing the IDF in the wake of the war.[20] They appeared to be threefold. First was the offensive option of crossing the canal. This was ruled out on political grounds. The second option was a withdrawal from the canal to the Mitla and Gidi passes into Sinai; this was rejected as a territorial concession. The third option was to adopt a defensive posture designed to prevent the Egyptian army from crossing the canal and achieving territorial gains. This made the canal a frontier.[21]

Since a defensive posture seemed most reasonable for political reasons, the tactical question was what kind of defense: static and stationary or mobile and flexible. The IDF, restricted to its political-territorial cul-de-sac, chose a defensive option in the Sinai that called for the establishment of fortifications that would become the Bar-Lev line.

Now the real debate began among the generals, all of them participants in the 1967 War. Generals Bar-Lev, Yeshayuh Gavish, and Avraham Adan (Bren) defended a stationary but well-fortified line. Generals Sharon and Tal were proponents of the mobile defense and were not convinced of the value of stationary fortifications next to the canal (*Maozim*). They argued that the Bar-Lev line only encouraged Egypt's War of Attrition. The fortifications signaled an Israeli presence on the canal for an unspecified length of time, which was an intolerable state of affairs for Egypt. "The Bar-Lev line therefore marked the continued preservation of the political, military, and territorial status quo, which the Egyptians sought so avidly to amend." [22]

The controversy among the generals over Israeli troop deployment after the start of the War of Attrition in 1969 was even more rancorous. It ended in a compromise that resulted in a second line of fortifications (*Taozim*) and the construction of a system of roads to serve the generals who advocated mobile defense. Somehow, the August 1970 cease-fire persuaded the IDF command that Israel had won the War of Attrition and, in the process, the Bar-Lev line strategy became Israeli doctrine in the Sinai. Meanwhile, the government troika supported by Yigal Allon objected to any withdrawal from the waterline except within the framework of a peace settlement. [23]

This was a case of wishful thinking and reaching for an impractical goal. The political strategy of Israel's cabinet actually became the greatest obstacle to negotiations, compromise, and peace. It was as if Israel were playing a one-handed chess game with itself even as Egypt was feverishly preparing to cross the canal.

While Egyptian preparations were moving along, a new debate erupted among the generals. David Elazar had become chief of staff in 1972, replacing Bar-Lev himself, and Sharon had been promoted to commander in chief of the Southern Command with Tal as Elazar's deputy; it appeared that the mobile-strategy generals were in command. Elazar felt that the first portion of the Bar-Lev line (*Maozim*) was not detrimental to "operational flexibility," [24] and he made almost no changes in the original conception, which had become entrenched. In effect, the *Maozim* were there to stay; the question that remained was how to deploy the forces.

Sharon and Tal did not defy the territorial concept either. They only objected to what they saw as a stationary defense as opposed to a mobile one, which they preferred. The line, argued Elazar, was not a defensive line, and in July of 1972 he closed a few *Maozim,* due to "Sharon's pressure." [25] But when Sharon retired to go into politics in July 1973, his successor, General Shmuel Gonen, sought to reopen the *Maozim.* But the onset of the war brought an end to the option of reopening the stationary outposts.

As Bar-Siman Tov concluded, "Just a short time prior to the outbreak of the Yom Kippur War, a compromise created a somewhat ambiguous conception. Despite the weakening of the permanent deployment along the waterline, no change had taken place in the mode of deployment in Sinai. On the event of the Yom Kippur War, neither of the two defense concepts was discernibly predominant. This lack of clarity was decidedly to Israel's disadvantage during the first few days of the war." [26]

The unresolved debate over the Bar-Lev line lingered into the "Generals' War" between 6 and 8 October. Changes in command were made: Bar-Lev replaced Gonen as commander in chief of the Southern Command and Bar-Lev finally resolved the debate in favor of Sharon and the Israeli crossing of the canal on 13 October.

The military results of 1973 clearly demonstrated the weakness of a rigid political-territorial doctrine that was not defensible even without the surprise element. The defense of the Sinai was possible only by a large army dedicated to the doctrines of defense, not an army indoctrinated in the offensive, in "the totality of the tank," as was the case with the IDF. Emanuel Wald's report, although perhaps too severe, blamed Ben Gurion for his failure to instruct the IDF in the basic Clausewitzian concept of the defensive.[27] It was Ben Gurion's concept of "transferring the war into the enemy's camp" that would make generations of IDF generals shun the defensive and embrace it reluctantly and ineffectively.[28]

In the course of the Yom Kippur War, three basic IDF doctrines were proven faulty and unsatisfactory. One was the concept of the territorial defense; another the total dependence on the offensive. The third was the concept of "the totality of the tank and armor" warfare espoused by General Israel Tal, the father of Israel's armored forces and its chief for seven years. Tal inculcated the spirit of the armored attack, the total tank, armored shock doctrines to his disciples, Gonen and Adan. His influence on Elazar, himself a former chief of an armored division, was considerable.

Yet, tanks cannot operate independently of other arms. Tank warfare must be integrated with artillery, armored infantry, and the engineering corps. The only way to defend a tank against antitank missiles is to integrate tanks with armored infantry, neutralizing the enemy's armored artillery antitank fire.[29] Artillery must be organically tied to tank and armored infantry to take over territory swept by tanks.[30] None of these ideas were taken into account by the IDF in its planning and training between 1967 and 1973.

The entire Sinai defense concept was flawed. Operation Dovecote (*Shovach Yonim*), a seventy-two-hour holding operation in the Sinai until the IDF could call up its reserves, was a plan which hoped that the regular armored forces could delay the Egyptian crossing and prevent the establishment of Egyptian enclaves.[31] Still another plan, *Sela* (or Rock), called for the addition of two armored divisions for a counteroffensive.[32] Neither plan worked. The IDF counteroffensive of 8 October saw three divisions repulsed by the Egyptians. The Agranat Commission, established to investigate what happened between 6–8 October, determined that both Dovecote and Rock were not successfully implemented.[33] Yet, as Bar-Siman Tov concludes, "The entire conception of the defense of Sinai, based on Shovach Yonim was insufficient to defend the Canal line." [34] Thus, even for *tactical* reasons, any Egyptian territorial aspiration was unacceptable. "The failure of the Yom Kippur War can, therefore, be attributed more to the conception of the defense of Sinai rather than to the Bar-Lev line, deriving from the politico-military conception that the territorial status quo could be protected by means of a small military force. This relied on the belief that Israel's deterrent capacity was adequate in order to ward off limited, as well as total war, and on the myth of the Bar-Lev line as an invincible fortification." [35]

The IDF did not perceive the serious changes Soviet training and Soviet military doctrine had wrought on the Egyptian army. Despite intelligence information concerning the creation of large Egyptian RPG infantry divisions armed with antitank missiles and the knowledge of the effectiveness of Soviet surface-to-air missile (SAM) batteries, the IDF command remained wedded to its own conception of warfare, ignoring what it knew was sure to come.

Placing the onus of responsibility on Israel's political leadership does not completely explain the IDF's behavior between 1969 and 1973, certainly not after 1972 when Israeli intelligence continuously reported on Egyptian preparations for war. The only question that could possibly remain in the minds of the generals was the date of attack. The Israeli security doctrine unfortunately had its political and military flaws. The 1967 War was won strategically but fought tactically again, thus providing another example that generals tend to fight the last war, not the next one.[36] Defense Minister Moshe Dayan had repeatedly warned that the Egyptians and the Arabs would not accept the dictate of 1967. The IDF intelligence had repeatedly sent warnings, especially after March 1973, that the Egyptians were preparing for war. But the IDF was hostage to the same notion held by the political elite.

The Egyptian feat of crossing the canal was a strategic success: Israel's territorial "arrogance" had to be modified at all costs. But the cost to Egypt was monumentally high, both in financial and manpower terms.

In an interview on 29 November 1977, in his Heliopolis headquarters, General Abdal-Ghani al-Gamasi told me that "the price we paid to cross the Canal of Shame was monumental indeed. In order to advance toward the canal, we have retreated from Egypt. Advancing into Sinai cost us years of economic development, of struggling against Egypt's ills: overpopulation, crowded urban areas, levels of education and jobs. The price was high but nevertheless necessary. It was inevitable."[37]

In Egypt, the Bar-Lev line was not only a strategic invitation, it was perceived in mythical terms. All of Egypt's military efforts and planning focused on crossing the canal. Once the Egyptians had settled on the eastern shore of the canal after 10 October, the IDF seized the initiative, successfully crossed the canal, and encircled the two Egyptian armies, which were saved from destruction by U.S. intervention.

The Egyptian plan was based on a Soviet model, whose basic tenet was that the best way to go to war is to train, prepare, and drill the army for full-time preparedness, to advance troops to the front, and to surprise the enemy strategically, throwing his defenses into disarray.[38] From 1968 onward, but especially after 1972, when General Saad al-Din al-Shazli was appointed chief of staff and the decision to go to war was made, most members of the Egyptian army knew little about the actual time and location of the attack. In the process, the Egyptians did manage to inflict serious casualties on the IDF.

Yet the army's fate depended totally on the element of strategic surprise. After October 10, the Egyptian army was at the mercy of the IDF. But, this time, as Dayan wrote, "the Egyptians did not run away." When General Shazli and his army tried to take the initiative on October 10 and move into Sinai, they failed miserably. Shazli's army made the attempt and lost 300 tanks in the process, costing Shazli his career. Shazli failed to realize that beyond the crossing of the canal, the plan to retake Sinai was doomed to failure and beyond the capacity of the Egyptian army.

On the whole, the results of the 1973 War were favorable for both sides. Israel learned to surrender territory for peace, with the help of the troop separation agreements with Egypt and Syria, conducted by Kissinger's brilliant diplomatic crisis management.[39] Egypt regained its military and political pride in the early successes of the war. For Egypt, the October War became the symbol of Egyptian national, political, and military pride. Above all, the 1973 War was the starting point in the long and difficult road that would lead to the eventual signing of the Egyptian-Israeli Peace Treaty in March 1979.

Paradoxically, both sides could call the war a victory. Had it not been for the fact that Egypt achieved psychological victory and strategic surprise, the process toward peace might never have begun. If the war had resulted in another devastating Egyptian defeat from start to finish, the Israeli-Egyptian Peace Treaty might never have materialized.

As we have seen, generals always tend to fight the last war, and Israel's generals were no exception. In this case, the same generals fought the same wars. The Israeli generals were their own exclusive judges, in a sense. They had won a smashing success in the 1967 War, they were in command in the War of Attrition, and they conducted the maneuvers and ran the study of results. But they remained tied to the "last" war, the 1967 War, and in fact, the last several wars.

The Egyptian lessons were easier and simpler to learn. They had lost in 1956 and 1967, and the policies of those wars had ended in defeat and disaster. Most of the military leaders of Nasser's generation had been purged, dismissed, or put on trial in the wake of 1967. The post-1967 Egyptian army was characterized by a new breed of officers who had been trained by the Soviets. More important, the political leader, in the person of Anwar Sadat, had changed, although it was Nasser who first uttered the cry of revanchism in the Sinai. Sadat was not tied to the failures of the past and proved cautious, patient, and more determined than Nasser. He was Truman to Nasser's F.D.R.

Israel did not voluntarily give up its territorial doctrine in the Sinai. It was forced to do so by a combination of Egyptian success, a Soviet threat, and American mediation which had promised territorial concessions. Today, Israel is once again willing to make territorial concessions to the Palestinians only because the latter have offered peace. But the Palestinians have no state, no real army, no territory, and no Soviet advisors. The Palestinian uprising certainly mars Israel's international image and creates dissension at home. But this is not the same as the high price of the 1973 *Mehdal.*

The psychology of the participants is also important in this case. Dayan was clearly aware that the situation on the canal was unacceptable to the Egyptians. Israel's intelligence services reported to the political leadership that the Soviets were training Egyptian troops. No Egyptian, Soviet, or Syrian activity was missed. And yet the Israelis were caught off guard. What was it that made even the doubter Dayan and others so confident? They could not conceive of the Egyptians crossing the canal by 10 October nor could they imagine them achieving what they achieved by 15 October without the Bar-Lev line. The need to change from an essentially offensive strategy into a defensive one was deemed "unacceptable." And they did not take into account the following:

1. The magnitude of the distress that would arise if the IDF, until now always victorious, were beaten even for two or three days.
2. The national mood if the Bar-Lev line were to be breached.
3. The force of the international community and the superpowers, especially the Soviet Union, which threatened to undo the Israeli achievement of crossing the canal and ousting the Syrians from the Golan Heights.

Israel did not err by underestimating the enemy or the threat. Rather, it overestimated its own capability. The IDF defensive strategy was proven inadequate. But how does one test such a strategy without experiencing its consequences? Unfortunately, the later writings of Generals Dayan and Tal demonstrate that the real lessons of 1973 still have not been learned. Many of the mistakes of the 1973 War were repeated in the 1982 war in Lebanon; in fact, the only serious lesson that was adapted was the use of deception and the element of surprise.

Territories are vulnerable assets. If you stay east of the canal, you can withdraw. For instance, if, in a preemptive act, Israel had crossed to the west on 5 October, it still would have had to face a massive Egyptian force equipped with RPGs and thousands of artillery guns and missile batteries, including SAMs, which would have resulted in prohibitive Israeli casualties.

In my view, Israeli political and military leaders failed to learn the real lessons of 1973. The defensive remains a dirty word with Israeli officers and generals, even though the lesson of 1973 was the shattering of the myth of Israeli invulnerability and invincibility. October 6–8 were the days of *Mehdal,* which means an absence of doing. The *Mehdal,* however, also describes the psychological impact on the nation, a kind of national trauma, as the inviolability of Israeli security was put to the test. Yet the only lessons that seemed to have been learned were technological—improvements in weapons, command, control, and coordination. In the 1982 Israeli invasion of Lebanon, there was no massive Soviet-trained army to contend with. The skies belonged to the Israeli air force. Yet a decisive victory still proved elusive. It was another fiasco.

Another lesson that was not learned concerns the process of decision making. The relationship between the IDF and the cabinet reveals a surprisingly primitive decision-making process. There was no policy planning staff before 1973, neither in the prime minister's office nor in the Foreign Office. There was no institutionalized decision-making process. Decision making was largely personal, conceptual, and nonanalytical. Writes M. K. Yehuda Ben-Meir: "National security, by definition, is a vital issue for any nation. For Israel, due to obvious considerations, this is doubly so. One would thus expect to find in Israel a highly developed decision-making process, using the most advanced tools available to modern-day decision makers . . . It is therefore more than surprising to find that Israel not only lacks basic tools of policy planning and analysis at the highest level of decision-making, not only falters at the highest level of elementary inter-departmental coordination, but that at the summit it has, in effect, *no organized and systematic decision-making process at all.*" [40]

Ben-Meir's study lists the following shortcomings or deficiencies in the present system:

1. Lack of a clear definition of the goals and aims of national security policy.
2. Lack of periodic reexamination of basic assumptions.
3. Lack of systematic staff work at the highest level.
4. Limited approach to problems. Each problem is dealt with individually, with an eye to specific and immediate goals. An overall view is lacking, as in an examination of each issue in a wider context.
5. Lack of overall planning.
6. Responsive and unplanned actions. As a result of the lack of policy

planning, decisions are made in reaction—sometimes hastily, even spontaneously—to immediate events, without thinking ahead and without proper examination of future consequences.[41]

As for the IDF, Ben-Meir makes this point:

A recurring theme in so many of the comments on and assessments of Israel's national security decision-making process is criticism of the disproportionate role played by the IDF in shaping Israel's strategy. Amos Perlmutter writes that "with the exception of the Soviet military, the IDF can be said to be the only military organization in the world that wields almost complete power over strategic and tactical questions . . . the intelligence, planning, and operational branches of the IDF, as well as the chief of staff, mold Israel's security doctrines. Rarely do civilian leaders make inroads into that decision-making process." Few observers would argue with this rather extreme diagnosis. Even more interesting and significant, criticism of the IDF's role in decision-making process is not limited to civilian circles, but is even more marked among former IDF generals and other high-ranking officers. Yariv states retrospectively that only the military possessed the staff requisite for strategy development. All the ministers, including the minister of defense, have had to rely on the military. Therefore, the military viewpoint has inevitably been influential. Avraham Tamir states: "What influences the decisions more than anything else are the papers of the General Staff, and they are tailor-made to the needs, the conceptions, and the viewpoints of the Army." [42]

The Agranat Commission Report of 1973 recommended the creation of a national security office in connection with the defense ministry. The only reform to be instituted was the establishment of AGAT (planning branch), a special IDF branch for planning and net assessment under the authority of the chief of staff. Its impact on the IDF was negligible. Later, under Defense Ministers Peres and Sharon, Tamir's agency became *Yallal* (the national security unit), which later came under the authority of the chief of staff and the ministry of defense. Unfortunately, under Sharon, it became his instrument for war in Lebanon and existed as a parallel command to the IDF command.[43]

No real superplanning agency was established in the wake of 1973. Israel's political elite does not want unelected bureaucratic or defense intellectuals intervening in national security decision making. The only large-scale security study and assessment ever made by the Israeli Knesset was made by Knesset member Dan Meridor of the Defense and Foreign Affairs Committee at the end of 1987. The Meridor Committee issued a secret report to the prime minister, the defense ministry, the foreign office, the IDF, and the intelligence community. Despite the report's severe criticism of Israel's security decision-making process, there was no public debate. Due to the disproportionate influence of the IDF and the intelligence community, there are no signs of security reforms in Israel in 1989.

For Egypt, its dedication to undo all the traces of what it perceived as aggression paid off handsomely. By 1977, Sadat realized that his army had no hope of ever again taking on Israel in a major war and winning. Instead, he cleverly flew to Jerusalem. Two years later, Sinai was once again Egyptian, and in the process he had gained an American ally, having ejected the Soviets from Egypt. Begin was free of the ideological burden of Sinai as a Jewish land, and Dayan, who had gone from being a national hero to a haunted leader, quickly and positively responded to Sadat's initiative. In fact, it was Dayan who offered all of Sinai to Sadat's emissary when they met secretly in Morocco. Sadat knew that at the end of a long series of protracted negotiations there would be the prospect of an Egyptian Sinai. To attain this goal, he would have to give up a cherished Arab form of hegemony, recognize Israel, and learn to live with it in peace. Most Israelis, except a militant right-wing minority, clearly understood the historic nature of what they could gain from giving up the Sinai.

7

Recovery from Defeat
The U.S. Army and Vietnam

Andrew F. Krepinevich, Jr.

The United States army that was ordered to South Vietnam in 1965 seemed better organized, trained, and equipped for combat than it had been at the outset of any of America's previous wars. Furthermore, its adversary was not the Kaiser's army, the Wehrmacht, the Japanese empire, or the Chinese People's Liberation army, but the black "pajama"-clad forces of a recently decolonized Third World nation.

Yet after eight years of fighting and watching more than 30,000 of its personnel be killed in action, the army completed the withdrawal of its forces without having defeated the enemy. Two years later, the South Vietnamese army it had trained and equipped for more than twenty years was overrun in a six-week blitzkrieg operation conducted by the Vietnamese Communists. If it was not evident in 1973, it was clear in 1975 that the army had lost its first war.

How has the army reacted to its defeat in the Vietnam War? According to studies of organizational behavior, the massive shock that defeat inflicts on the organizational psyche significantly enhances the prospects for reform and change.[1] Simply stated, defeat in war provides great incentives for a military organization—even one like the army, which had previously enjoyed great success practicing its methods of warfare—to change these methods lest it repeat its failures in the next war.

This chapter explores what happens when a military organization that has experienced defeat fails to acknowledge its share of the responsibility for that defeat. The discussion that follows consists of three parts: how well the army prepared itself for the revolutionary war (or People's War, as the Communists referred to it) that it encountered in Vietnam; how the army fought the war in Vietnam; and how the army views its role in the war and how this view affects today's army.

The Army and Revolutionary Warfare

The army that exists today and that existed during the Vietnam War was forged during the Second World War—particularly in the campaigns in Western Europe—and, to a lesser extent, during the Korean War. It is an army that, quite naturally, viewed the Soviet Union as its primary adversary and Central Europe as the decisive theater of operations. Consequently, the army oriented its doctrinal development, force structure, and field training heavily toward a high-intensity, conventional conflict. In limited conflicts like the Korean War, this conventional orientation was supplemented by the need to minimize U.S. casualties.

Despite this preoccupation with conventional warfare on the plains of Central Europe (or perhaps because of it), the Cold War era saw the focus of conflict shifting increasingly to the Third World. Many new and unstable nations were emerging from the breakup of the old empires of Europe. Over the past forty-five years the army has engaged in numerous military operations and campaigns but, ironically, none of any significance have occurred on the Continent.

In fact, the first decade of the Cold War saw the army involved in helping not only Korea but friendly nations like the Philippines and Greece combat insurgencies that threatened their survival. Washington dispatched Military Assistance Advisory Groups (MAAGs), comprised primarily of army personnel, to aid the Filipino and Greek governments in their battles against Communist insurgents. The MAAGs, however, found their greatest utility as conduits for providing material support to the host regimes, as opposed to delivering tactical or strategic advice on how to defeat the insurgents.[2]

From 1950 onwards the army also found itself becoming increasingly a part of the French effort to subdue the Viet Minh insurgents in Indochina. Following the French defeat at Dien Bien Phu in May of 1954 and their subsequent withdrawal from Indochina in the wake of the Geneva Accords that summer, the United States increased its involvement in the region. Consequently, the U.S. MAAG in Saigon assumed the responsibility for equipping and training an army for the newborn Republic of Vietnam.

The assignment of Lieutenant General Samuel "Hanging Sam" Williams to head up the MAAG reflected the organization's increased status and responsibility.

Lieutenant General Williams and his staff generally ignored the French experience in battling the Viet Minh insurgents. The Vietnamese Communists had proven themselves expert practitioners of a form of "unconventional warfare" (the army's term) they called People's War. Nevertheless, Lieutenant General Williams favored a more traditional, conventional approach to the conflict in Indochina. With the approval of the U.S. Joint Chiefs of Staff, Williams focused his efforts on organizing the Army of the Republic of Vietnam (ARVN) to be capable of fighting a Korea-style conventional war.[3] His premise seemed to be that if the North Vietnamese attacked, they would try to copy their unsuccessful Communist cousins in Pyongyang rather than attempting to repeat their recent successes against the French.

Over the next four years the MAAG organized the ARVN along the same lines as its American counterpart. By the time Lieutenant General Williams departed in 1959 he could claim to have accomplished his mission: building a South Vietnamese army capable of conducting corps-sized operations.[4] Unfortunately for the army and its South Vietnamese counterparts, the North Vietnamese and their South Vietnamese compatriots—the Viet Cong—had no intention of deviating from the People's War strategy.

The strategy of People's War recognized the Vietnamese Communists' inability to win quickly, or through conventional means, a war of national liberation against an advanced western military power.[5] Recognizing their dramatic inferiority in conventional military power, the Vietnamese Communists sought, through People's War, to involve the enemy in a protracted struggle conducted methodically and designed to obtain a series of intermediate objectives that would lead over time to the overthrow of the Saigon regime.

The People's War strategy consists of three phases. In the first phase of contention, the insurgents build their political infrastructure, espouse a popular cause to recruit membership, and, perhaps, conduct selected terrorist acts against the regime. When the insurgents accumulate sufficient strength, they move on to the second phase, that of overt violence. This phase continues all operations initiated in the first phase but is also characterized by guerrilla operations against political, economic, and military targets. These operations are designed to gain resources for the insurgents while denying them to the government, and to convince the people—by persuasion or intimidation—to support the insurgents' efforts to topple the existing order. Phase three, the counteroffensive, occurs when the military balance has finally tipped in the insurgents' favor as a consequence of the two earlier phases of their campaign. In this phase the insurgent guerrilla forces are combined into large formations to contest openly the regime's conventional or quasi-conventional military operations, as well as by the operations conducted in the earlier two phases.

There are two keys to the growth of the insurgency: an administrative weakness on the government's part that allows the insurgents the freedom to organize and a popular cause that enables the revolutionaries to attract a following. The insurgents must maintain access to, and eventual control over, the population. This access guarantees the supply of manpower, food, medicine, and intelligence needed for insurgent operations to continue and expand. At the same time, this control over the population denies these same resources to the government. If the insurgents succeed in the critical task of winning the "hearts and minds" of the people, they can gradually sap the government's strength while increasing their own, building toward the final counteroffensive phase.

Attempting to use conventional means of warfare to combat an insurgent movement employing a People's War strategy will prove frustrating at best and disastrous at worst. The offensive campaign in counterinsurgency warfare is pacification—denying the insurgent access to the population—and not the traditional conventional mission of closing with and destroying enemy main forces. Insurgent logistical support runs in the opposite direction—from the "front" to the rear—found in conventional wars. Thus counterinsurgency campaigns that have as their primary focus the destruction of enemy forces and supplies through conventional methods, with little or no effort devoted to pacification, will achieve only temporary success, as the insurgent will replenish his manpower and supplies by drawing them from the unprotected population.

In 1960, both the continued growth of Viet Cong insurgency warfare against the Saigon regime and the election of John Kennedy as president of the United States brought into sharp focus the army's inclination to treat the conflict in Vietnam as a variant of its Korean War experience. Kennedy entered office only two weeks after Soviet leader Nikita Khrushchev had given a major speech voicing his support for Mao Zedong's efforts to spread communism in the Third World through so-called wars of national liberation. Kennedy was determined that the United States would play an active role in helping friendly nations defeat this form of aggression.

Robert McNamara, Kennedy's secretary of defense, designated the army as the Defense Department's executive agent for counterinsurgency warfare. The president's problem was how to engineer a "revolution from above" and refocus the army to attain a more balanced approach to the postwar conflict environment.[6] In his address to the graduating class at West Point in 1962 Kennedy warned the army that it faced "another type of war, new in its intensity, ancient in its origins—war by guerrillas, subversives, insurgents, assassins; war by ambush instead of by combat; by infiltration instead of aggression, seeking victory by eroding and exhausting the enemy instead of engaging him. It requires . . . a whole new kind of strategy, a wholly different kind of force, and therefore a wholly different kind of military training."

The president took a personal interest in the development of a national counterinsurgency capability, which he saw as a key element of his strategic doctrine of flexible response. Kennedy ordered the creation of a Special Interdepartmental Group (Counterinsurgency) to oversee the process and recalled to active duty former army chief of staff, General Maxwell Taylor, to head the group and serve as his personal military assistant. During his first year in office, Kennedy also sought and won from Congress approval for a huge expansion in the army's Special Forces, popularly referred to as the Green Berets. The Special Forces had been established in 1952 to provide a capability for unconventional warfare. The army was also directed to develop a doctrine and force structure—which was to include the retraining of its conventional forces—to provide the nation with an effective counterinsurgency capability.

The army leadership's reaction to these directives from the White House was distinctly negative. General George Decker, then the army chief of staff, told the president that "any good soldier can handle guerrillas." General Lyman Lemnitzer, the chairman of the Joint Chiefs of Staff, remarked to a friend that Kennedy was "oversold" on the importance of guerrilla warfare. General Earle Wheeler, soon to succeed Decker as the army's chief of staff, stated that "the essence of the problem in Vietnam is military," not the "hearts and minds" focus of counterinsurgency. Even Maxwell Taylor, brought into Kennedy's inner circle of advisors because of his reputation as an intellectual with an open mind, instructed the president privately in the Oval Office that "we good soldiers don't have to worry about special situations . . . Any well-trained organization can shift the tempo to that which might be required in this kind of situation."[7] In summary, the army leadership's position was that if you could fight and win the big war in Europe, then you could certainly win a little war against guerrillas in Southeast Asia as well.

The army leadership generally viewed Kennedy's revolution from above as a "fad" foisted on it by the "New Frontier" crowd, a collection of "Whiz Kid" analysts and armchair generals who did not understand the nature of war the way they, the professionals, did. Thus their approach to Kennedy's directives was mechanistic in nature. The army would "answer the mail," but with little enthusiasm or initiative. The counterinsurgency mission would be accorded considerable lip service, but few resources. Numerous conferences were organized, study groups were convened, analyses were conducted, and directives were issued— all designed to address the problem and insure that all soldiers were trained in counterinsurgency operations. However, counterinsurgency never succeeded in working its way into army doctrine or force structure, despite Kennedy's interest and the growing U.S. commitment in Vietnam.

The army failed to develop a coherent doctrine for counterinsurgency through its educational system, field manuals, military journals, or field training. For example, much of the mandated "increase" in counterinsurgency instruction at army schools was effected through minor modifications to standard instruction, which was then recertified as "counterinsurgency-related."[8] Progress in providing field manuals on counterinsurgency was uneven at best. By 1965, on the eve of the army's deployment of combat forces to Vietnam, the service's Combat Developments Command reported that "doctrine for the organization, employment, and support of an advisory organization, other than the special forces, does not exist."[9]

More informal means of doctrinal development and dissemination also failed to bear fruit. Army journals like *Military Review* and *Army* retained their overwhelming focus on conventional war. The few articles that did address counterinsurgency often served more to stake out a particular branch's role in counterinsurgency warfare or to demonstrate how minor modifications to conventional doctrine would prove effective in combating insurgents.[10] Little thought was given to the fundamental differences between conventional and revolutionary war.

Furthermore, most of the field training conducted involved only minor modifications to conventional doctrine. The traditional mission of closing with and destroying the enemy—in this case guerrillas—remained the centerpiece of training and served as a precursor to the search-and-destroy tactics employed in Vietnam.

The army did little to structure its forces for counterinsurgency warfare. When people think of army units in Vietnam, they usually picture airmobile and the Special Forces. Ironically, these forces were not developed, nor were they deployed to Vietnam, with counterinsurgency in mind. During the 1961–65 period, the army was preoccupied with eliminating the Pentomic division structure it had adopted during the mid-1950s to function more effectively on President Eisenhower's massive retaliation (that is, nuclear) battlefield. These divisions were reorganized into so-called ROAD (Reorganization Objective Army Division) units, designed for high-intensity conventional conflict in Europe in line with Kennedy's new strategic doctrine of flexible response. Indeed, the army was so anxious to reorganize for conventional war that it had drawn up plans to replace the Pentomic divisions with ROADs even before Eisenhower had left office.

There was no similar push by the army to adopt airmobile forces or increase its Special Forces. As noted, President Kennedy mandated the increase in Special Forces. These elite counterguerrilla forces achieved some notable successes in pacification while under the CIA's aegis in Vietnam from 1962–63. Once the Green Berets were brought under the MACV's (Military Assistance Command, Vietnam) control, however, their mission was quickly changed to focus on more conventional operations: attempting to seal South Vietnam's borders from infiltration and supporting regular units in their attacks on Viet Cong war zones.[11]

Airmobile forces were originally conceived to help the army "survive" the focus on nuclear warfare during the Eisenhower administration. It was felt that airmobile divisions that could rapidly concentrate and disperse would fit nicely into the president's "New Look" for defense, while protecting the army from the budgetcutter's scalpel.[12] Unfortunately for the army's small but influential "Airmobile Mafia," the divisions cost an estimated $3 billion apiece, pricing them out of the administration's frugal defense budget.

The airmobile proponents, men like Generals Hamilton Howze, Robert Williams, and Harry Kinnard, advanced their proposals again when the Kennedy administration came into office. Again, their focus was on fielding air assault divisions and brigades in Europe to increase mobility and supplement the Air Force's close air support capabilities. McNamara was willing to consider the airmobile concept, and testing began in 1962 with the creation of a provisional unit, the Eleventh Air Assault Division. Over the next two years, the division underwent extensive testing, culminating in the Air Assault I and II exercises.

Despite the significant increase in the use of helicopters in Vietnam beginning in 1962, the Eleventh Air Assault Division tests focused almost exclusively on the European conflict environment.[13] Indeed, one evening in November 1964 after the last field exercises had been completed, General Johnson, the army chief of staff, came down from the Pentagon to be briefed by the generals who had overseen the evaluations. Johnson asked the generals, "How well do you think an airmobile division would do in conflicts such as Vietnam?" The generals responded that they had not really thought about it and that their entire focus had been on justifying the airmobile concept in terms of its potential role in conventional conflicts.[14]

The Army in Vietnam

Thus the army was ill-prepared, both doctrinally and in terms of its force structure, for counterinsurgency warfare when, on 7 June 1965, General William Westmoreland, the commander of the MACV, requested that forty-four "Free World" maneuver battalions be sent to South Vietnam to fight the communist insurgency that seemed about to overwhelm the ARVN and seize control of the country.[15] In his request, Westmoreland cited intelligence that elements of two North Vietnamese army divisions were believed to be in the Central Highlands or poised to enter that region from Laos. This information, combined with the increased activity of the Viet Cong main force units, convinced Westmoreland that the insurgency was moving into its third phase and that only U.S. conventional forces could redress the balance.

Westmoreland's concept of operations was to employ this infusion of U.S. military power to seize the initiative from the enemy by the end of 1965. Twenty-four additional battalions would be required for follow-up operations from 1966 to mid-1968. These operations would focus on attacks against the enemy's base camps and sanctuaries. By mid-1968, sustained ground combat operations would mop up the last guerrilla forces and push them across the border.[16] All in all, a very conventional approach to an unconventional war.

The conventional mind-set is also revealed in the army force planning process in Saigon and Washington, which centered around achieving the 3 to 1 ratio in maneuver battalions thought necessary for successful operations in conventional war. As calculated by MACV, the request for forty-four battalions provided Westmoreland with a 3.2 to 1 advantage over the Communists.[17] The addition of twenty-four battalions, it was believed, would offset North Vietnamese infiltration and maintain the 3 to 1 ratio.

By determining force requirements in this manner, MACV omitted some 100,000 Viet Cong irregulars who played a central role in the communist strategy of People's War, although not part of the main force units. Worse still, Westmoreland and his planners had failed to address what MACV would do in the event the enemy retreated in the face of superior conventional military power, from phase three operations back to phase two guerrilla warfare operations. The rule-of-thumb force requirement for combating an insurgent force was not 3 to 1, but 10–15 to 1.

After the Ia Drang Valley Campaign in October–November 1965, in which the U.S. First Cavalry Division (airmobile) defeated elements of two North Vietnamese army divisions that stood and fought, the Communists reverted to guerrilla operations. Denied an opportunity to encounter and destroy the enemy in a battle of annihilation, General Westmoreland opted for a strategy of attrition, whereby the war would be won by destroying enemy forces faster than they could replace them. This approach fit comfortably within the army's conventional view of its task as seeking out and destroying the enemy. In South Vietnam, these search-and-destroy operations were designed to identify and eliminate the enemy's main force units and their logistical base camps. But this strategy neglected the enemy's underlying strength: access to the population. As long as MACV remained focused on trying to fight a "big-unit war," the Communists could always replenish both their manpower and supply losses through levies on the South Vietnamese people.

It quickly became obvious that the success of the attrition strategy depended on the ability to reach the so-called crossover point, that point at which communist units were being depleted faster than they could be replenished. Having chosen to ignore the indirect approach of attrition through pacification—that is, cutting off the supply of manpower at its source—MACV had to seek combat to make its strategy work.

Ironically, by failing to place its major effort on pacification, MACV subverted its own attrition-through-combat approach by failing to threaten the only target the Viet Cong could be forced to fight for: the people. Consequently, from 1965 through the end of 1967 the Communists were able to revert to guerrilla-style attacks against allied forces, while refusing to fight unless the terms of combat were in their favor. Battalion-sized attacks by the Viet Cong and North Vietnamese army decreased from 9.7 per month in the last quarter of 1965 to 1.3 per month in the last quarter of 1966. Meanwhile, the number of small-unit enemy attacks increased by 150 percent over the same period.[18] MACV, however, was not to be budged from its conventional mind-set. The army continued to search for the big battles that the enemy had no intention of fighting. Despite the reversion of the conflict to guerrilla war, General Westmoreland launched numerous large-scale operations, such as Operations Attleboro, Cedar Falls, and Junction City. Concerning Operation Junction City, an attack into War Zone C during February 1967, Westmoreland observed that "the operation employed for the first time all our different types of combat forces, including paratroopers and large armored and mechanized units."[19]

It is both the supreme irony and a testament to the depth of the army's conventional mind-set that, as the enemy was clearly moving away from anything resembling conventional warfare, the army was moving toward it. At the time Westmoreland's forces were trying to create the "body counts" that would enable the allies to reach the crossover point, MACV data revealed that the enemy was initiating 88 percent of all engagements.[20] Thus the Communists held the initiative. They could refuse battle or accept it on their terms and thus regulate their casualties and frustrate Westmoreland's attrition strategy.

By Autumn 1966 senior U.S. defense officials were beginning to have serious doubts concerning MACV's ability to reach the crossover point. President Johnson expressed these doubts personally to General Westmoreland during the latter's visit to Washington in April 1967.[21] Yet MACV's only alternative to attrition was a proposal to invade Laos across the panhandle region from the demilitarized zone (DMZ) to the town of Savannakhet on the Mekong River.[22]

In January 1968, as the United States approached the third anniversary of its commitment of ground combat forces, the war had reached a military stalemate: Westmoreland had failed to reach the crossover point, but the Communists had been denied the victory that had seemed just within reach in the Spring of 1965. At this point the Communists launched the Tet Offensive which, although a tactical military defeat for Hanoi, proved to be a decisive victory in the battle of political will.

Although Hanoi failed to spark a mass uprising throughout South Vietnam, it did succeed in temporarily derailing the promising pacification program that MACV had finally initiated in 1967 under civilian leadership. Furthermore, the Communists also killed more than 1,000 Americans, part of their war of attrition on U.S. will, and increased the refugee population, and its burden on the Saigon regime, by more than 800,000 people.[23]

The attacks sparked a major policy review in Washington. The Joint Chiefs' advice to newly appointed Secretary of Defense Clark Clifford was to pursue more of the same attrition strategy for an indefinite period. Even if Clifford chose this course, however, the Joint Chiefs could not give him any assurance of success. The civilian leadership, and President Johnson's "Wise Men"—a group of distinguished senior officials—refused to endorse the continuation of a high-cost strategy that had such a low potential payoff.[24] The president sided with his advisors and against the military. He withdrew from the 1968 presidential race, froze U.S. troop strength in Vietnam, and gave the go-ahead for the "Vietnamization" of the war. The U.S. military was no longer asked to produce victory, but rather to effect a withdrawal of its forces in a manner that would allow the ARVN a fighting chance against its internal and external foes. This it accomplished (arguably) by 1973.

The Army after Vietnam

After the war, the army's initial reaction to its experience in Vietnam was one of avoidance. During much of the 1970s, the army sought, not to learn the lessons of the war, but to forget the ordeal. The service adopted the nation's "No More Vietnams" mood with a vengeance. Instruction in counterinsurgency at the army's War College and Command and Staff College, never great, effectively disappeared. Army leaders viewed the 1973 Arab-Israeli War as an opportunity to get the officer corps back to the basics of conventional war.[25] When in 1976 the army reissued *Field Manual 100–5*, its bible on how to fight, deleted were the three chapters on "sublimited war" included during the Kennedy era.

The army's force structure also changed dramatically. Airmobile forces were refocused on Europe as part of the Tricap Division concept. When the concept failed to pan out, the 1st Cavalry Division (airmobile) was converted to an armored division. The Special Forces were radically reduced in size.

Toward the end of the decade, however, the army entered a period of introspection. Between 1976, when General Westmoreland's memoirs were published, and the early 1980s, when they were joined by Colonel Harry Summers's analysis of the war and, later, General Bruce Palmer's memoirs, a picture emerged of what the service felt were the lessons of the war.[26] Summers's monograph, *On Strategy,* was particularly well received within the army. Copies were sent to all army general officers and to the White House, and the work became standard reading at the army's senior service schools. In his monograph, Summers presents four contentions that are strongly at odds with the evidence on the war but that conform to the army's conventional mind-set.

1. The Vietnam War was best understood as a conventional war.
2. The army was engrossed, rather than repelled, by the Kennedy administration's emphasis on counterinsurgency.
3. MACV's strategy focused too heavily on the internal, guerrilla, threat to South Vietnam and too little on the external threat posed by North Vietnam.
4. The United States should have adopted a strategy similar to the one employed in the Korean War; that is, U.S. forces should have created a barrier along the DMZ, through Laos to Savannahkhet along the Mekong River, while letting the ARVN destroy the Viet Cong.[27]

Thus Summers places the blame for the army's failure in Vietnam on the service's purported unwillingness to be true to its conventional war mind-set.

Summers paints for the army a comforting picture of the service being led astray by the civilian leadership: first being mesmerized by Kennedy's emphasis on counterinsurgency, and then being betrayed by Lyndon Johnson's failure to mobilize popular support for the war by declaring war on North Vietnam and calling up the reserves. Finally, the civilians are blamed for not giving the military a clear objective to achieve.

In summary, Summers tells the army that it was right all along: the war should have been fought the tried-and-true conventional way. There is nothing to learn from Vietnam, except perhaps to exact some preconditions from the civilian leadership should the army ever find itself in this type of conflict in the future. Indeed, these preconditions were worked into Secretary of Defense Caspar Weinberger's so-called Six Tests that should be met prior to committing U.S. forces to future Third World conflicts.[28] They were seconded by President Reagan's admonition that the United States should not commit its combat forces in Third World conflicts "unless we are prepared to let them win." The implication is that, although the U.S. armed forces were perfectly capable of winning the Vietnam War, they were somehow prevented from doing so by the civilian leadership.

As the army entered the 1980s, then, it did not view the U.S. loss of the war as primarily attributable to its failure to operate effectively in Vietnam. Furthermore, the statements of the commander in chief and his secretary of defense supported the army's belief that what needed to be changed was not army strategy and doctrine, but the commitment of the American political leadership and the American people to the army's way of war. Unlike President Kennedy and Secretary of Defense McNamara, President Reagan and Secretary of Defense Weinberger would not challenge the army to prepare for a "new type of war." Rather, they would reinforce the service's view that, since Vietnam was not the army's defeat, there was no need to change the service's approach to future unconventional conflicts.

Thus the army of the 1980s proved very similar to the army of the early 1960s in its approach to what is now referred to as "low-intensity conflict." By the 1980s, the United States had recovered from its Vietnam "syndrome" of the 1970s. Growing instability in Central America, the 1979 oil shock, and the Iranian seizure of the U.S. embassy in Teheran led Americans to realize that U.S. interests in the Third World had not evaporated with the nation's defeat in Vietnam.

In 1980 they elected to office an activist president willing to use force to protect U.S. interests, albeit in a more constrained manner than his early Cold War predecessors. Indeed, under Reagan, U.S. policy, as it had under Kennedy, favored the use of security assistance programs to help friendly states threatened by insurgency, although intervention by U.S. combat forces was not ruled out.[29] How did the army prepare for this new era of American activism in Third World conflicts? Not surprisingly, it did so in a manner very similar to the way it approached the conflict in Vietnam, as a review of army doctrine, force structure, and involvement in such conflicts makes clear.

Army doctrine continued to be mired by the service's poor understanding of unconventional war. Simply defining low-intensity conflict (LIC) took the Pentagon eighteen months. Revealingly, the services have not found it necessary to define either mid- (conventional) or high-intensity (conventional/nuclear) conflict. Adding to the confusion, the Pentagon later chose to distinguish between "conflict" and "war."

The army's field manual 100–5, *Operations*, has been published twice since 1981. Yet the decade passed without a revision being made to field manual 100–20, *Low-Intensity Conflict*, which was prepared during the Carter administration and is generally considered to be a poor manual. Field manual 100–5 does not address LIC at all, other than to make the point that the army's AirLand Battle doctrine—its doctrine concerning conventional war in Europe—"applies equally to the military operations characteristic of low-intensity war." [30] The army was preparing to publish a much-improved version of field manual 100–20 in 1990. Yet the problem of LIC exclusion from the service's "mainstream" operations manual will maintain unconventional warfare's second-class status in army thinking.

The low-level emphasis on LIC doctrine in general, and counterinsurgency doctrine in particular, is found in the army's lower-order field manuals. For example, field manual 90–4, *Air Assault Operations*, an eighty-five-page manual, devotes only eleven lines of text to counterinsurgency. Field manual 7–30, *Infantry, Airborne, and Air Assault Brigade Operations*, is devoted to describing how these brigades fight, yet there is no reference to unconventional operations. Even field manual 31–22, *Command, Control, and Support of Special Forces Operations*, focuses on how the Green Berets can support large-scale conventional military operations.[31] The manual is currently being reworked to conform better to the AirLand Battle doctrine.

Further evidence of the conventional war mind-set is found in the Army Command and General Staff College's field circular 71–101, *Light Infantry Division Operations*, where overwhelming emphasis is placed on the light infantry division's (LID) operating as part of a larger force conducting conventional operations. The Army Command and General Staff College's scenario for LIDs, designed to "help clarify emerging doctrinal principles which explain how to fight . . . in a contingency environment," finds the LID opposing a Soviet motorized rifle division. Guerrilla forces are included as part of the scenario, but the students involved in the exercise are instructed to keep counterguerrilla operations at a minimum and focus on conventional operations.[32]

In summary, as field manual 7–70, *Light Infantry Platoon/Squad Operations*, observes, army "doctrine seldom changes and when it does, it occurs slowly." The statement provides an apt description of army doctrine concerning LIC over the past thirty-five years.

Instruction at the army's service schools continues to reflect its conventional war orientation. At the senior service colleges, where emphasis is placed on the operational art and strategy, the teaching of LIC is accorded low priority. For example, at the Command and General Staff College, the LIC faculty continue to struggle for an allotment of 30 hours out of 600 total hours of instructional time.[33] And this allotment covers all types of LIC—counterterrorism and so-called peacetime contingency and peacekeeping operations, as well as counterinsurgency. At the army's War College, the core curriculum devotes a mere two days to the study of the Vietnam War.[34]

Nor is LIC training focused on unconventional conflict. The most publicized LID exercises at the army's National Training Center, for example, have been those in which light infantry battalions were attached to heavy brigades for conventional operations. Joint exercises conducted in Central America in 1986–87 were primarily concerned with preparing U.S. forces to fight Nicaragua's Sandinista army as opposed to any one of a number of guerrilla forces in the area. In the Fall of 1986, the army's 82nd Airborne Division engaged in its most extensive exercise since the Vietnam War. During training in Honduras, the division defeated the "aggressor" force through conventional tactics. When the "aggressors" reverted to guerrilla warfare, however, the division commenced raids akin to search-and-destroy operations against suspected guerrilla positions. Similar exercises occurred in February and May of 1987.[35]

As in the early 1960s, the army of the 1980s gave the appearance of developing forces capable of effectively conducting counterinsurgency operations. The force buildup, however, was not conducted with LIC in mind. For example, the recreation of several Special Forces Groups still found them training primarily to fight as guerrillas, as opposed to fighting guerrilla forces.

The army's LIDs, touted as the service's regular force contribution to the LIC mission, were, in fact, created primarily to preserve the army's ground force primacy against the Marines in a Persian Gulf contingency. The LID's primary requirement was an ability to deploy rapidly using less than 500 C-141 transport sorties. Its putative capability in LIC was considered a "bonus" by force planners. In fact, the division is poorly structured, especially in its support elements, to conduct unconventional operations against forces employing guerrilla tactics. Thus the LID, like the airmobile division of the 1960s, was designed primarily for the army's preferred conventional war conflict environment.

Finally, those units that would provide crucial support in counterinsurgency operations—the army's psychological operations and civil affairs units—have over 90 percent of their forces in the reserves. This could pose problems if U.S. forces are committed to protracted conflict, given the demonstrated reluctance of U.S. political leaders to call up reserve formations. The result may be units being committed piecemeal, or operating at reduced strength and without key support elements.

Given this post-Vietnam approach to insurgency warfare, how is the army performing in its security assistance role? "American Military Policy in Small Wars," a study by four army colonels who were senior fellows at Harvard University, reveals great similarities between today's army and the army of the early 1960s.[36]

The colonels examined the case of U.S. security assistance to El Salvador. They found that U.S. advisors sent to El Salvador, like their counterparts assigned to Vietnam a generation ago, were not considered to be on the army's "fast track" to early promotion. Rather, these officers are from the army's "Third Team." These junior field-grade officers (majors and lieutenant colonels) are men who generally have not attended the Command and General Staff College—a key "ticket" for advancement in the service. Few go on to command battalions once they complete their tour in El Salvador. The position of MILGROUP commander, the head of the advisory team, has proven to be difficult to fill at times since the officer corps views the position as "non-career enhancing." One officer recalled being offered the position after five others had refused it. On another occasion, the army ended up asking a retired colonel to return to active duty to fill the position.

The Salvadoran armed forces were just as ill-trained and poorly organized to combat insurgents as the ARVN was. This was due in significant part to the support provided by the U.S. military, including the army. Salvadoran army officers trained in the United States at the army's General Command and Staff College, or at the army's Infantry School at Fort Benning, Georgia, were exposed to the army's conventional approach to waging war. It was not surprising, therefore, that Salvadoran military operations were reminiscent of those conducted by the army in Vietnam. For example, the colonels' study revealed that the Salvadoran forces used the fifty-four howitzers provided by the United States primarily for "harassment and interdiction" fire. Thus the targets were not known guerrilla positions, but rather suspected enemy locations. This violates the counterinsurgency principle that calls for extreme care in the use of firepower. In a war in which the insurgents are mixed with the people—the supposed goal of the pacification effort—the slavish and indiscriminate use of firepower could have deleterious consequences, as was shown in Vietnam.

The Salvadorans also tried to mimic their American mentors in the area of mobility. Salvadoran troops were loaded down with equipment and preferred to travel by truck or helicopter. Thus, in conventional military terms, they were mobile. To the extent that rapid reaction forces are required, this is a useful attribute. They were not "ground mobile" like the lightly equipped insurgents, however, who could travel cross-country and mix with the people far more effectively than can the government troops on the highway or in the sky.

Just as their American counterparts proved unable to adapt to a change in Vietnamese communist strategy after the Ia Drang Valley Campaign, so too has the Salvadoran military had difficulty adapting to the Marxist insurgents' reversion, after defeat in several "final offensives" prior to 1984, to more of a phase two posture of protracted guerrilla warfare. Yet the Salvadorans still preferred to operate in battalion formations, employing trucks and helicopters for movement and depending heavily on artillery and close air fire support.

As in Vietnam, scant attention was paid to psychological warfare, whereas the pacification campaign received much publicity but little sustained support, either from the United States or from the Salvadorans themselves. San Salvador's poor pacification record can hardly be laid at the feet of the United States army. However, the Salvadoran army's performance is reminiscent of the ARVN and the army's approach in South Vietnam before the civilian-dominated Civil Operations and Revolutionary Development Support staff assumed control of pacification in MACV in 1967.[37] The much-publicized Salvadoran National Campaign Plan, while impressive on paper, suffered from a military failure in execution. The same lack of understanding and impatience that led ARVN units—in Operation Sunrise and the Strategic Hamlets and Hop Tac pacification programs—to terminate quickly their pacification effort in an area and move on in search of main-force guerrilla bands, plagued the Salvadoran army as well.[38] In summary, the Salvadoran army, beneficiary of the American army's military assistance and Vietnam experience, attempted to fight an insurgent force in much the same manner as its South Vietnamese counterparts a quarter of a century earlier.

Conclusions

Has the United States army recovered from the Vietnam War? In many ways, it has. Today's soldiers are brighter and better equipped than those of a generation ago. They are better prepared to fight on short notice and for a more sustained period than their Vietnam era counterparts. In sum, they are better able to perform the mission they have traditionally been prepared for: to wage conventional war. This was clearly confirmed during the Persian Gulf War in 1991.

The army is not significantly better able to cope with insurgency warfare, however. Ironically, as the threat of a major war in Europe fades and a post–Cold War era emerges, insurgencies and other forms of internal or unconventional conflict seem destined to endure as a plague on a troubled Third World. The army seems to feel it has turned the corner in preparing for LIC counterinsurgencies through its successful operations in Grenada and Panama. In both instances, however, the service was conducting what are referred to as "peacetime contingency operations." These operations, which were relatively low in firepower intensity, are not significantly different from conventional military operations designed to seize and occupy territory. The fact is, the army is still woefully deficient in its ability to cope with insurgencies and other unconventional wars.

There are those who would say that this begs the question, that the fact is the American public will not permit any more Vietnams. Certainly the "Six Tests" enunciated by Secretary Weinberger are designed to avoid U.S. participation in Third World conflicts. And yet, public pronouncements and popular opinion are often poor barometers on which to base military contingency planning. The army found itself fighting a war in Korea less than six months after Secretary of State Dean Acheson declared Korea outside the United States' defense perimeter. It was President Kennedy who said of the South Vietnamese, "It's their war. We can help them . . . but in the final analysis, it's their people and their government who have to win or lose this struggle," and President Johnson who echoed the theme in 1964 when he said, "We are not about to send American boys nine or ten thousand miles from home to do what Asian boys ought to be doing for themselves."[39] Yet less than a year later, the army found itself fighting in Vietnam. And despite Weinberger's "Six Tests," the army found itself invading Grenada (hardly a "vital" United States interest), and the U.S. forces were dispatched on their ill-fated mission to Beirut, and, more recently, to Somalia (arguably without a clear objective and certainly without the means to "win").

Successive administrations have stated that security assistance is the preferred option for providing U.S. help to friendly regimes threatened by insurgencies, and that direct intervention would occur only as a last resort, after all other means have been exhausted. As shown above, this policy does not eliminate the army's need to prepare for this type of war any more than the United States' desire to avoid war in the Persian Gulf eliminates the need to prepare for conventional conflict.

Given the recent events in Europe and the fact that insurgency wars are the prevalent form of conflict in the Third World, it would seem prudent for the army to study revolutionary war, train for it, and allocate significant resources for it. Yet the army has not done this. Consequently, the service runs the risk of perpetuating its inability to help effectively counter revolutionary warfare movements. Given the growing focus on the Third World as a source of markets and raw materials, this could be a shortcoming of significant consequence for U.S. economic and security interests.

8

China's Recovery from Defeat in 1979

Gerald Segal

The title of this chapter may be considered controversial. Many would not accept that China was defeated in the brief border war with Vietnam in February 1979. And even among those who would accept that China was defeated, many would suggest that China is far from recovered from the debacle. In contrast to these points of view, this chapter will argue not only that China was defeated in 1979, but that, in the past decade, it has more than recovered. Indeed, there were considerable political forces in China eager to take advantage of the defeat in 1979 in order to shape a very different China.

The key to the analysis that follows is the definition of defeat as the inability to achieve objectives.[1] This definition is especially important in the case of the 1979 war because China initiated the hostilities and therefore can be deemed to have had initial objectives that were more than a mere reaction to the actions of others. Indeed, it will be argued that China was eventually able to achieve most of its initial objectives, but only after a decade of major reforms in domestic and foreign policies. Victory, therefore, came only after military defeat and only through nonmilitary measures. In this case, war was the failure of policy by other means.

On 17 February 1979 Chinese forces attacked Vietnam along their shared border. China was responding to some recent events—notably the Vietnamese defeat of Chinese-backed Khmer Rouge forces who had previously controlled Cambodia. But China had also grown increasingly concerned with Vietnam's "tilt" toward the Soviet side of the Sino-Soviet conflict, and consequently China had sought support for its attack on Vietnam from the United States, Japan, and other Western powers.

While these and other motives will be discussed in greater depth below, some mention needs to be made of China's past practice of using force beyond its frontiers. Despite a remarkably good reputation in the West for not being an aggressive military power—at least since the 1970s —China has used force in major engagements more often than any other great power since 1949.[2] China has certainly lost more soldiers in combat since 1949 than any other great power. There were not only major wars in Korea (1950–53) and India (1962), but other major engagements included various offshore island crises, the commitment of antiaircraft forces to the war against the United States in Vietnam in the 1960s, and of course the border conflicts with the Soviet Union. Although it has been common for Western analysts to explain Chinese action as careful uses of forces intended to deter enemy threats, the reality is that China was often aggressive in its definition of what it viewed as defensive activity, and certainly made a practice of trying to teach its neighbors lessons, including India in 1962 and even the Soviet Union in 1969.

Thus China's resort to force and even its willingness to bear an extraordinary level of casualties (some 20,000 dead in three weeks of fighting in 1979) are far from unique in recent Chinese history. It is also true that in a number of these engagements, China failed to obtain its objectives and therefore could be said to have been defeated. The most serious case of defeat was clearly the border clashes with the Soviet Union in 1969, but as they say, that is another matter.

The war in 1979 had a distinctive mix of Chinese motives.[3] First, China explicitly claimed it wished to "teach Vietnam a lesson" that it could not use force with impunity in East Asia. This not-so-veiled declaration by China was, like the attack on India in 1962, intended to shore up China's claim to be the leading power in Asia. In 1962, China was able to send a convincing message to the world, not to mention to India and other Asians, that China was the dominant power. In 1979 China had recently emerged from the Cultural Revolution, and with the death of Mao in 1976 and the eventual consolidation of power by Deng Xiaoping in 1978, it felt itself more capable of reasserting regional leadership after years of internal self-obsession.

Second, and somewhat less explicitly, China intended to force Vietnam to withdraw from Cambodia. The calculation seemed to be that if China could so scare Vietnam into believing that it could not defend its northern frontier, Vietnam would see it could not sustain its occupation of Cambodia. China and Vietnam had supported the Khmer Rouge when they seized power in 1975. But the Khmer Rouge grew increasingly radical in both foreign and domestic policy, leading to border clashes with Vietnam. The Vietnamese, now in need of assistance in postwar reconstruction, had been turning more to the Soviet Union for that aid. Relations with China deteriorated because of the already tense Sino-Soviet relationship.

Indeed, the third Chinese objective was to send a message that the Soviet Union could not protect its allies in East Asia and therefore demonstrate that the Soviet Union was a "paper tiger." Other states who might be contemplating seeking closer relations with the Soviet Union would see there was little purpose in doing so and therefore China's own position would be enhanced in its rivalry with the Soviet Union. China had discovered, in the 1971 war between its ally Pakistan and its rival India, that India was able to deter Chinese pressure in support of Pakistan by calling on the Soviet Union to threaten China's northern frontiers. In the 1979 war, China was trying to send a message that it could stand up to Soviet pressure.

Finally, China could also be said to have been interested in stopping the incidents along the Sino-Vietnamese frontier which were distracting at a time when China claimed it wanted to focus on economic reconstruction. In fact, this, the least important of China's objectives, was the one Peking cited most often in support of its attack on Vietnam in February 1979.

Yet China achieved none of these four primary objectives and was forced to withdraw its troops after three weeks of fighting. One provincial capital was taken and much destruction was wrecked on the Vietnamese border areas. But the operation took far longer than planned, and the casualties were far heavier. All of China's objectives were in a sense psychological, in that the specific targets on the ground were less important than how fast and devastatingly they were seized. This was a didactic war and, as any good teacher knows, the method can be as important as the message. In this case, without an effective military operation, the message was not conveyed.

China did not teach Vietnam a lesson as much as Vietnam taught China a lesson—the same lesson it had taught the Americans before them: do not take on Vietnamese communists on their own territory. There was little that the Chinese could do to the Vietnamese that the Americans had not already done on a much greater scale and for much longer.

Neither did China drive the Vietnamese out of Cambodia. Vietnamese border units were able to take on the Chinese regular army and inflict sufficient casualties to make the Chinese feel the pain even more sharply than the Vietnamese. If one accepts that China never had any intention of marching on Hanoi, then it was merely sufficient for Vietnam to slow the Chinese. Victory came from a draw because the initiator needed to win in a big way to win at all.

Of course, if Vietnam was not scared into pulling troops out of Cambodia, this was in part because of the careful Soviet response. As in the 1971 Indo-Pakistan crisis, all the Soviet Union would have had to do was to make sufficient noises about Chinese vulnerability along the Sino-Soviet frontier. This would have alerted China to the need to retain a strong presence along the northern frontier and to be swift in dealing with Vietnam to prevent the Soviet Union from planning a counterattack in defense of Vietnam. The Soviet Union did not have to send vast amounts of supplies or even take military action against China in order to be seen as acting as a reliable ally. Like the United States in various Arab-Israeli wars, there is a huge advantage in crisis management when you are backing the winning side. To deter is far easier than to compel.

Finally, the incidents along the Sino-Vietnamese frontier did not cease after the 1979 war. Indeed, China's attack so stirred up emotions that the border area became more dangerous, and casualties in the ensuing years were far higher than they had been before 1979. China achieved nothing, not even its most defensive objectives, and the reasons can be divided into three categories: military problems, domestic priorities, and foreign policy distortions.

Military Problems

The narrow specialist will seek the specific reasons for China's defeat in the state of its armed forces. Although this is only a small part of the explanation, it is still important. Because so much of the apparent Chinese strategy depended on a swift and successful attack on Vietnam, the effectiveness of the People's Liberation Army (PLA) was a necessary condition for victory.[4] This was a tough assignment, because victory had to be so swift that it scared Vietnam and yet did not allow time for the Soviet Union to either send supplies or take major military action along the Sino-Soviet frontier.

Some might even suggest that this was an impossible task and that, in effect, the problem was one of grand strategy rather than of its cousin, military strategy. Certainly the pressing problem concerned the specific military tactics. It was already apparent before 1979 that the PLA was, unlike in 1962, in no shape to fight such a swift, surgical war. Ever since the early 1960s, the professionalism of the PLA had been neglected, and no major provision of new equipment had been undertaken. In the late 1970s after Mao's death, the PLA had been debating what new policies should be adopted for modernization, and it was well known that the armed forces were not ready for a major operation. Deng Xiaoping himself, by 1979 China's paramount leader, had led the drive for modernizing the PLA and had overseen its first tentative efforts.

Although Deng should have known that the PLA was not up to the task, he may not have fully appreciated just how far behind the PLA had fallen. After all, Deng's own record as a military strategist is far from stunning,[5] and he, like many leaders before him, could have been deceived by hubris and distracted by other pressing objectives. Likewise, it is hard to believe that the PLA professionals would have been very keen for combat given their military unpreparedness. The more devious among them might have felt that defeat in a limited war would support their case for additional funding and modernization. They certainly had been unable to win the debates about professionalism before the war, but they clearly had their hand strengthened after defeat. On balance, it seems most likely that the decision to attack Vietnam was political and that the PLA followed orders. As was evident in the Soviet invasion of Afghanistan, military professionals do not always make the decisions they either should or could.[6]

Domestic Priorities

As has already been suggested, the PLA was given an unreasonable task considering its previous neglect of professional training and modernization and its need for new equipment. In the late 1970s Chinese leaders were emerging from the chaos of the Cultural Revolution and the post-Mao succession struggle. There was much uncertainty about the priorities of domestic policy and the specific types of reforms to be undertaken. Not until the end of 1978 was Deng Xiaoping able to take relatively firm control from Mao's designated successor, Hua Guofeng. The four modernizations were begun in earnest in December 1978, but this was only the beginning of the process. As we know from most recent events in China, the reforms were always fluid and debates were constant.[7]

Deng decided to attack Vietnam at the same time he was consolidating power in domestic political and economic affairs. He might have been looking for a swift victory to speed his campaign along. Or his judgment might have been distorted by a sense of overconfidence. But he certainly had yet to sort out his domestic priorities. Military modernization was one of the four modernizations but was not yet clearly relegated to fourth place. Although a wise leader might have recognized what Deng eventually came to accept—that a peaceful international environment was essential for domestic reconstruction—in late 1978 Deng still thought he could have both an active defense policy and a growing economy.

The reforms that followed the defeat in 1979 gave priority to economics—growth was the goal. Debates continued about the pace of growth, and Deng was instrumental in scaling back some of Hua's more extravagant claims concerning what might be achieved in a short space of time. But even Deng had to learn that didactic wars were the "luxury" of the rich and the already powerful. In 1979, China was neither, and the war made both objectives harder to achieve.

Foreign Policy Distortions

As we have already suggested, foreign policy considerations were a key element of China's motives for going to war. But because China possessed such a distorted view of the outside world and of the specific threats to its security, it was almost bound to have had the troubles it did.[8] Several key foreign policy problems can be identified.

First, China was excessively hostile toward the Soviet Union.[9] This led the Chinese to worry that because Vietnam was relying on Soviet aid after the defeat of the United States in 1975, it meant that Vietnam was a growing threat to Chinese interests. During the war against the United States, both China and the Soviet Union had supported Vietnam, but China was unable and unwilling to provide the reconstruction aid that Vietnam needed after 1975. China had also made clear its desire to dominate East Asia and its unwillingness to tolerate an independent-minded Vietnam. Where else should the Vietnamese have turned in order to deter China?

As a result, China had an excessively hostile view of Soviet foreign policy. China instigated the 1969 incidents and then refused Soviet offers of talks. China also denounced East-West detente and refused to join any arms control negotiations. And China joined with the United States in denouncing virtually every aspect of Soviet foreign policy.

Clearly China was leaning too far toward the United States in the erroneous perception that the Soviet Union was an active threat to Chinese security. If China had not attacked Vietnam, or if Peking's ally, the Khmer Rouge, had not been allowed to antagonize the Vietnamese, then China would not have experienced nearly the same level of Soviet hostility. Indeed, when Deng Xiaoping went off to the United States and Japan in early 1979 seeking support for his war, the Western powers were far less anti-Soviet than the Chinese. Peking was expecting too much from countries that had grown used to managing their relations with the Soviet Union and retaining at least a modicum of detente. If China was expecting the United States to counter Soviet pressure on China in support of Vietnam, then Deng badly misjudged the state of international relations.

Indeed, Peking generally had a far too confident and China-centered view of its place in international relations. This overconfidence, combined with the 1979 mood of domestic uncertainty, led to the bullyboy tactics of China's attack on Vietnam. This, in turn, led to China's embarrassing defeat and the need to modernize both China's domestic and foreign policies.

In the decade following March 1979, China went a long way toward full recovery from defeat. In May 1989 and at the Sino-Soviet summit, China was more secure from external threat than at any time in the past several hundred years. Sadly, by 4 June Peking's attitude toward the outside world, not to mention the world's view of China, would change yet again. But the reality remained basically the same; China had recovered from defeat to be more prosperous, secure, and successful than even the most optimistic might have dreamed in 1979. The recovery can be explored through the same three categories as the defeat—military issues, domestic politics and economics, and foreign policy.

Military Reforms

As has already been suggested, the PLA was put in an impossible position in 1979. Its professionals knew before the war started that, given the previous neglect of professional concerns, these same professional issues would have to form the focus of any recovery program. The faults were known within the army before the fire of combat illuminated them for all to see.

The problem for military reformers was how to achieve their objectives while spending less money.[10] In the past, the perceived (erroneous) solution was to enhance "fighting spirit" and especially its political components. In the so-called "red versus expert" equation, the stress was on being a revolutionary. But when the matching, if not the more intense, Vietnamese fighting spirit was set against that of the PLA, China needed something more to win. Vietnamese forces were also combat-hardened and the fighting was on their home ground.

The war was so costly to China that, although defense spending rose sharply as a percentage of the total government budget to cover the war expenses, far less money was available for the much tougher process of rebuilding the PLA. But given the abysmal state of professional standards in the PLA, much could still be done to improve the armed forces' effectiveness, even without spending money.

Indeed, many have argued that the most significant weaknesses in the PLA were in those areas where money was not the issue.[11] Professionalism had suffered when political purity was ranked higher than professional competence. Thus it was relatively easy, and inexpensive, to bring in younger, better educated recruits and train them in more professional skills. Officers were sent for training in new colleges. All soldiers spent more time on the training field, and they were rewarded for accuracy and efficiency, rather than for political orientation. Simulators were acquired, and more time was spent in real maneuvers.

Contacts were also cultivated with foreign armed forces in order to learn what China had been missing. Given the strategic environment of the time, most of these contacts were with Western professionals. Military officials were sent around the world to learn about new technologies, obtain them where possible by legal means, but steal them if necessary. Foreigners were invited into China to advise on training, managing, and, of course, handling new weapons.

But the swiftest and most far-reaching reforms involved the intangible aspects. Professionalism was primarily an attitude rather than a piece of hardware. This new spirit, when added to modern doctrine, helped the PLA learn to fight wars for the last quarter of the twentieth century. These changes meant learning to rely less on massed manpower and more on handling modern weapons. Active defense, including the holding of fortified or valuable positions, could be more likely. Instead of moving at the pace of foot-slogging infantry, men would be made more mobile by putting more people on vehicles. Combined-arms operations would require coordination with air force and even naval forces where possible. Most of this new doctrine was first developed for a potential war with the Soviet Union, but as foreign policy priorities shifted, so did the focus shift to more limited war, with greater emphasis on the navy and less on the ground forces.

These revised doctrines and greater professionalism, combined with a new foreign policy (to be explored below), made possible perhaps the most significant feature of the money-saving modernization—the reduction in size of the PLA by one million men in 1985–87. Nearly half of these soldiers were transferred to units outside the PLA that did the same tasks. By reducing the PLA by one quarter, more money was available to be spent on those who were left and needed new weapons.

What is more, these cuts in personnel came at a time when the absolute amount of the PLA defense budget was increasing, even though it was a smaller slice of the total state budget. Overall growth in the economy was meeting some PLA demands, much as the original theory of the four modernizations suggested it would.

To make matters even rosier for the PLA, defense industries were "privatized" or at least turned over to civilian industry. Thus the legendary inefficiencies were transferred out of the PLA budget. Some firms were kept within PLA control and made to make a profit by providing goods for the booming civilian economy. No precise figures are available for these reforms, but it is clear that all these changes meant there was more money per soldier in the new, more professional PLA. Special efforts were also made to cut waste and come to grips with corruption. The slimming down of the PLA, and especially the removal of many senior officers whose interests were less than professional, meant some progress was possible in this realm. Later in the 1980s, however, as corruption became more rampant in society as a whole, it infected the PLA as well.

Funds were also raised from arms sales abroad—a new venture for a China used to selling at "friendship prices" to revolutionary colleagues in the developing world.[12] With the fortuitous expansion in the Iran-Iraq war from 1980, China soon developed a lucrative export market for its rugged, second-rate hardware. China was able not only to earn money from direct sales but also to expand its range of contacts with various states in the Middle East who provided new technology and expertise on weapons development. Perhaps most paradoxical of all was Israel, who had no diplomatic relations with China, but did have much expertise in reverse-engineering Soviet-type arms and in dealing with Third World arms markets. This covert relationship also allowed China to obtain technology and assistance in Israel, therefore avoiding the expense of buying weapons off Western shelves.

But it was China's direct sales policy in the Iran-Iraq war that attracted the most attention. Not only did China sell weapons to all sides, but it quickly became clear China was playing in the major league of tank, fighter, and even missile exports. This was not a matter of furtive loads of AK-47s to a band of rebels in the hills; this was a policy of high-profile convoys of key weapons in a major set-piece war. By the late 1980s, China had become the fourth largest arms exporter; in terms of volume rather than value, it was even ranked third in some weapons. China had become a superpower, at least in its ability to deliver weapons and aid to allies at long distance during combat. Its missiles sunk Western-flagged tankers, drawing the Soviet Union and Western powers into the hot war. Later Chinese sales of intermediate range ballistic missiles to Saudi Arabia fueled an arms race and fed regional paranoias. By selling arms to all sides, Peking, and most important of all, the PLA, made nearly a billion dollars in each of the peak years in the late 1980s.

Back home, China was not merely buying more "obsolete" weapons for the PLA. She was far more sensibly increasing spending on research and development of a new generation of weapons that made the best of smaller-scale technology. The trick was often merely to add a new gun to an old platform, as on tanks, or to add new avionics to tried and tested aircraft. The key was funding for research and development and help with the technology from Western and even Israeli specialists. By the end of the 1980s, every service arm of the PLA had received major new items of equipment, with the navy and air force doing especially well. China was still not acquiring large numbers of weapons in each category, but they were making sensible decisions about new directions for a modern force instead of merely filling the old arsenals with shinier kit.

Of course, no soldier will claim to be happy with the level of defense spending or the state of professionalism. Although reports stressed the major improvements in PLA professionalism, by 1988 there were persistent reports that the PLA was not happy with the still shrinking percentage of state spending.[13] In 1989, and especially in 1990, the PLA's share of the

budget increased, although inflation was affecting the real spending figures. By 1993, and on the back of a booming economy, defense spending
had clearly increased, effectively doubling its total since 1989.

Perhaps the best test of the state of the PLA was the same one that
showed up its problems—war. In March 1988 China launched a brief but
bloody attack on Vietnamese forces in the Spratly Islands in the South
China Sea. This was not a massive engagement and really only involved
naval forces, but it demonstrated that China was able to use its most modern equipment in an effective operation against precisely the country that
had so humiliated it in 1979. Although this minor skirmish is far from a
perfect test of the PLA's recovery, the 1988 operation was as good a symbol
as one was likely to get.

Domestic Priorities

Military modernization, as important as it was to improving the fighting strength of the PLA, was merely part of a broader move toward professionalism in China that underpinned the four modernizations. The other
three modernizations—in economics, society, and science and technology—were far more important as national priorities and established
the foundation for changes in the PLA. Without greater stress on general
education and skills, the PLA could not have required greater professionalism for its own forces.[14]

Indeed, professionalism was the distinctive way in which Deng Xiaoping and his fellow reformers saw the issue of political reform. They were
never enamored of Western-style pluralism, but preferred to divide the
duties of state and Party to allow professionals to get on with their jobs.
Of course the emphasis on professionalism inevitably left less scope for
commissars to intervene in technical matters, and the tensions between
the old forces of "Red" and "expert" persisted.[15]

The professionalism was energized by a willingness to open the system up to outside influences. This essential feature of reform linked
domestic and foreign policies and meant greater de facto pluralism in
domestic politics. Regions, industries, and individuals were given more
freedom to develop their own skills and profits. Market forces were introduced to a larger extent than ever before, but the Party never surrendered
its ultimate control. As long as the economic reforms were producing the
spectacular growth rates of ten plus percentage points, the dissent could

be bought off. The "nothing succeeds like success" principle was working in a way that led professionals to believe they were set on a course for a far better life.

We cannot explore all the details of the economic reforms here, but we can suggest that what was happening in the PLA was part of a much broader trend in the society at large. A new generation of leaders were emerging with a greater commitment to professional values. Although China was still ruled by a seemingly unchanging gerontocracy, expectations at the lower and middle levels of authority were changing. Reforms were successful, but they were also dangerous. The PLA, as the chief law-and-order authority, although benefiting from increased professionalism and a growing economy, also saw the risks the more lucrative economy might bring: greater instability and the greater attraction for young people to work in the civilian economy and stay out of the PLA. Trouble was brewing in civilian-military relations.[16]

Yet these domestic reforms were clearly key components of China's recovery from defeat. Had China merely modernized its military, at considerable cost to the society at large, it might have seen the instability of military rule typical of Latin American countries. China made a wise choice to stress the civilian basis of recovery and expect that the military professionals would be satisfied with what they obtained from the larger and expanding pie. China's recovery was primarily political, economic, and social rather than military. Only after economic progress had ground to a halt in the last half of 1988 and after social crises began to develop into major political ones did the PLA find its professionalism under the most severe test of all: in June 1989 it was called upon to fire on its own people. Had the PLA not swiftly returned to the barracks in 1990, PLA professionalism might have been under far greater risk than it already was. As it turned out, the pursuit of professionalism in the PLA was given strong support in the early 1990s as the economy stabilized and produced impressive growth. And the growing economy made possible an increase in defense spending, much as the professionals had been promised.

Foreign Policy Reforms

The Chinese did not recognize the problems in their foreign policy in 1979, or if they did they did not let on. It seems to be in the nature of Chinese foreign policy (and perhaps of most authoritarian regimes) that changes in policy are rarely acknowledged as anything except changes made by others. But it remains true that China made major changes in its foreign policy in the decade following the war, all of which contributed to a more open and independent position.

In 1979 China had "normalized" relations with the United States by establishing full diplomatic relations. This was clearly part of a Chinese attempt to build an anti-Soviet coalition and was soon followed by some of the most extreme statements of Chinese support for anti-Soviet positions. If normal relations between great powers actually mean some balance of competition and coexistence, then China's excessive tilt to the American side of the East-West conflict was abnormal. Certainly by the standards that China was to establish by 1982, when a more officially independent line was taken, the period between 1979 and 1982 was not normal.

The previous error in China's policy toward the United States was an exaggerated belief that the United States could both control Soviet behavior and offer trade and technology benefits to China.[17] Both features of the Chinese assessment were to change, but only gradually. The sharp tilt to the American side did not stop Vietnam from consolidating power in Cambodia or prevent the Soviet Union from invading Afghanistan. When President Reagan came to power in 1981, it became clear that the United States would take a tougher line against the Soviet Union, but President Reagan was also more favorably disposed toward such anticommunists as Taiwan. Thus the United States was both making life more difficult for the Soviet Union, and thereby relieving part of what China perceived to be a defense burden, and also making life more difficult for China. What is more, China saw that many of the trade benefits it sought from the United States could actually be obtained more easily from American allies, especially in Japan and Western Europe. China had made a mistake in focusing too narrowly on the United States, although it was true that detente with Washington made it more possible to do business with some American allies, most notably Japan. The lesson for China was the need for greater independence, which was easier to achieve at a time of superpower conflict.

China also saw that the Soviet Union was not as powerful as it once thought. The Afghan rebels were able to tie down the Soviet superpower and the Soviet economy seemed to be grinding to a halt. What is more, China itself was reassessing the meaning of socialism as part of its domestic reforms and gradually came to realize that the Soviet Union was not as nasty as it had appeared when China was defining ideological purity more narrowly.[18] Indeed it can be argued that the most important and basic steps along the long road to Sino-Soviet detente were taken in China before 1982.

Of course, the brief tenure of Yuri Andropov as Soviet leader in 1982–83 speeded up the process of detente, just as the Chernenko phase meant a pause in detente. The coming to power of Gorbachev in 1985 clearly moved Sino-Soviet detente along at a faster clip. In general, most of the improvement can be traced to bilateral concessions and the interconnection of the domestic reform process in both states. China's concentration on the so-called "three obstacles"—Soviet occupation of Afghanistan, Soviet support for Vietnamese occupation of Cambodia, and Sino-Soviet tension along the frontier—were eased by Soviet concessions on all three. But China also gave ground. A summit meeting was agreed upon even while a pro-Soviet regime ruled in Kabul, there was a pro-Vietnamese one in Phnom Penh, and the bulk of Soviet troops remained along the frontier. Detente was a matter of mutual concession and flexibility.

From the perspective of our special concern with China's recovery from defeat in 1979, it is striking that China found it could best push Vietnam around by doing nearly the precise opposite of what it did in 1979. The best way to get Vietnamese troops out of Cambodia was to tease Vietnamese-backers in Moscow with the dream of detente with China in exchange for pressure on Vietnam. Gorbachev, anxious to reduce drains on the Soviet economy and drags on Soviet diplomacy, was amenable to striking a deal. The Soviet Union could improve relations with China, the Association of Southeast Asian Nations, and East Asia generally, if it would "encourage" Vietnam to leave Cambodia and open up to reform and foreign trade. By 1986 there were already clear signs that the Soviet Union was exerting significant pressure on Vietnam. Although Vietnam did not completely withdraw from Cambodia until September 1989, by 1988 China was satisfied that the Soviet Union was exerting the right kind of pressure.

Perhaps the most striking evidence for China was the March 1988 Spratly clash, when the Soviet navy in the region declined to deter China from taking on the Vietnamese. Vietnam was still nominally more of a Soviet ally than China was, but Gorbachev knew that Deng would take it very badly if Soviet forces risked war with China for the sake of Vietnam. The contrast between this inactivity and Soviet deterrence of China in 1979 could not have been more striking. China had found that the way to show Vietnam who was the most important regional power was effectively to remove the Soviet Union and the United States from the equation and thereby leave Vietnam open to China's mercy. To that extent, the declining influence of both superpowers, and especially the Chinese ability to get on with both of them at the same time, was vital to China's recovery as a great power in the region. Getting the politics right was far more important to China's success than any modernization of the PLA.

By 1989, when Vietnamese troops had left Cambodia, the story of China's recovery from defeat could be said to be complete. Yet another Chinese objective was achieved by virtue of detente with Moscow. Similarly, Vietnam became far more solicitous of detente with China, and border incidents came to a virtual halt. The cessation of hostilities along the border had more to do with China's less aggressive posture, but Vietnam's more conciliatory mood was certainly influenced by the pressure that came from Sino-Soviet detente.

In fact, by 1990 Chinese recovery became part of an entirely new state of international affairs in the region. The normalization of Sino-Soviet relations achieved at the Peking summit of Soviet and Chinese leaders in May 1989, along with major strategic changes in Eastern Europe later in the year, meant that Soviet power was declining everywhere. The Soviet Union made clear its intention to withdraw entirely from Vietnamese bases, and even the United States was discussing cutting its forces in East Asia. As the superpower overlay was pulled back from the region, China could see some prospect of attaining regional hegemony, at least in military terms. Japan was the only uncertainty in this scenario, but then, militarily at least, Japan was especially vulnerable. Not since the Western imperial powers' arrival in East Asia several hundred years ago had China been more able to throw its weight around in the region. China was more secure than at any point in these several hundred years, but it was also more of a threat to its neighbors than at any time in that period.

The evidence from the specific Chinese case is clear. Recovery from defeat is primarily a matter of politics and grand strategy rather than of military reform or the acquisition of new technology. To be sure, China's recovery did include these more narrow reforms, but they were all part of a broader domestic political reform. China still has not admitted that it made mistakes, although it may be a matter of waiting for the key decision makers to die before the truth will become known, as happened in the Soviet Union after Brezhnev's death.

But Deng urges us to "seek truth from facts," and the facts of reform and the consequent recovery in Chinese policies are clear. China not only reformed its armed forces, but it opened up its politics as a whole to professional interests and market forces as part of the process of putting economics in command. In foreign policy, excessive dependence on the United States was replaced by more practical relations with all states, including those who had no diplomatic relations with China. Peking also became less hostile to the Soviet Union and eventually normalized relations with Moscow. This reform, more than any other shift in foreign policy, allowed China to meet its main objectives of 1979. Sadly for the Chinese, it took nearly a decade to fully realize the folly of their ways and implement reforms.

Although it is difficult to draw lessons for other countries, the Chinese case does seem to uncover certain features of the process of recovery that may be useful in other case studies.

1. It is not necessary for there to be new leaders in the state concerned, but new policies are certainly needed.
2. The specific military strength of the armed forces is not the main cause of defeat; it is instead the error of the politicians who send the army into battle.
3. Getting grand strategy right is essential, but it is not sufficient to ensure a full recovery. Sound domestic reforms and some narrow military reforms are also essential.
4. Because the problems leading to defeat seem less subject to quick military fixes, recovery will be lengthy; no short-cuts exist.
5. No state seems capable of recovering solely on its own, and a more genuinely independent policy and effective recovery seems to depend on better relations with a wide range of states. The paradox seems to be that independence requires more interdependence.

For students of Chinese affairs, it will not be surprising that recovery from defeat should be essentially a political process. It was Mao Zedong who left us with the Clausewitzian notion that political power grows out of the barrel of a gun. Mao also noted that the Party must always control the gun, but long-term and stable control of that gun depends on using it wisely. The war in 1979 was not a wise use of force because the political calculations were so off the mark. Had Mao been watching his successors, he might have added that military power really grows out of the intelligence of smarter politicians. And even military power is no substitute for wise political judgment, as the tragic events of June 1989 made plain.

Notes

Introduction

    1. The case study technique allows military historians from divergent backgrounds to venture into the relatively neglected field of comparative history. A group effort can treat more facets of a significant issue than would be possible for a single historian, thus increasing the breadth of coverage and, presumably, the analytical power of the collection. Many important collections of case studies have been published in recent years. Some of the best include Paul M. Kennedy (ed.), *Grand Strategies in War and Peace* (New Haven: Yale University Press, 1991); Brian Bond (ed.), *Fallen Stars: Eleven Studies of Twentieth Century Military Disasters* (London: Brassey's, 1991); Allan R. Millett and Williamson Murray (eds.), *Military Effectiveness* (Boston: Allen and Unwin, 1988); and Charles E. Heller and William A. Stofft (eds.), *America's First Battles, 1776–1965* (Lawrence: University Press of Kansas, 1986).

    2. The study of overextension and retrenchment as well as an exploration of the lessons to be learned from the experiences of the great powers have been the focus of much scholarly attention in recent years: see, for example, Paul Kennedy, *The Rise and Fall of the Great Powers* (New York: Random House, 1987); Michael Doyle, *Empires* (Ithaca, N.Y.: Cornell University Press, 1986); and Jack Snyder, *Myths of Empire* (Ithaca, N.Y.: Cornell University Press, 1991).

    3. It is beyond the scope of this introduction to revisit the debate between classical and scientific approaches to the study of international relations. For useful discussions, see Hedley Bull, "International Theory: The Case for a Classical Approach," *World Politics* 18 (1966): 361–77; Klaus Knorr and James Rosenau (eds.), *Contending Approaches to International Politics* (Princeton, N.J.: Princeton University Press, 1969); Stanley Hoffmann, *Gulliver's Troubles or the Setting of American Foreign Policy* (New York: McGraw-Hill, 1968); and Robert Jervis, Richard Ned Lebow, and Janice Gross Stein, *Psychology and Deterrence* (Baltimore, Md.: Johns Hopkins University Press, 1985).

    4. Edward A. Kolodziej, *French International Policy under De Gaulle and Pompidou: The Politics of Grandeur* (Ithaca, N.Y.: Cornell University Press, 1974);

and Lawrence Freedman, *The Evolution of Nuclear Strategy* (London: Macmillan, 1985), esp. 313–24.

5. As Liddell Hart has argued, "It is essential to conduct war with constant regard to the peace you desire" (Basil Henry Liddell Hart, *Strategy* [New York, 1974], 353; quoted in Paul Kennedy [ed.], *Grand Strategies*, 2).

Chapter 1

1. Quoted in Piers Mackesy, *The War for America, 1775–1783* (Cambridge, Mass.: Harvard University Press, 1964), 288.

2. J. K. Laughton, *Letters and Papers of Charles, Lord Barham Admiral of the Red Squadron 1758–1813* (London: Navy Records Society, 1907–11), 2: viii; W. M. James, *The British Navy in Adversity* (London: Longmans, Green, 1926), 16–18; John A. Tilley, *The British Navy and the American Revolution* (Columbia: University of South Carolina Press, 1987), 277.

3. R. G. Albion, *Forest and Seapower: The Timber Problem of the Royal Navy, 1652–1862* (Cambridge, Mass.: Harvard University Press, 1926), esp. ch. 7; but see R. J. B. Knight, "New England Forests and British Seapower: Albion Revised," *American Neptune* 46 (1986): 221–29.

4. Jonathan R. Dull, *The French Navy and American Independence: A Study of Arms and Diplomacy, 1774–1787* (Princeton, N.J.: Princeton University Press, 1975); H. M. Scott, *British Foreign Policy in the Age of the American Revolution* (Oxford: Oxford University Press, 1990).

5. David Syrett, *The Royal Navy in American Waters, 1775–1783* (London: Scolar Press, 1989); Nicholas Tracy, *Navies, Deterrence and American Independence: Britain and Seapower in the 1760s and 1770s* (Vancouver: University of British Columbia, 1988); N. A. M. Rodger, *The Insatiable Earl: A Life of John Montagu, 4th Earl of Sandwich* (London: Collins, 1993), 212–300; Michael Duffy, *Soldiers, Sugar and Seapower: The British Expeditions to the West Indies and the War against Revolutionary France* (Oxford: Oxford University Press, 1987), 3–37.

6. A manpower study covering the whole of the military and naval effort during this war is badly needed. Dull cites the lack of skilled manpower as a considerable restraint on the effectiveness of the French fleet (*French Navy*, 286–87). Advantage would seem to lie with the French at the start of the war, with the effect of the *Inscription*, but the reserves ran short by the closing years—a parallel to the shipbuilding effort. This is the view of Martine Acerra and Jean Meyer, *Marines et Revolution* (Rennes: Ouest France, 1988), 31. It is clear that the British effort to raise men started very slowly, with the government very wary, because of the political unpopularity of the war, of an early full-scale impressment.

7. Daniel A. Baugh, "Why Did Britain Lose Command of the Sea during the War for America?" in Jeremy Black and Philip Woodfine (eds.), *The British Navy and the Use of Naval Power in the Eighteenth Century* (Leicester: Leicester University Press, 1988), 149–69. Baugh also makes the point (163) that the British ultimately had far more skilled seamen than France, and certainly more than Spain.

8. See Jonathan R. Dull, "Mahan, Sea Power, and the War for American Independence," *International History Review* 10 (1988): 66.

9. This is the argument put forward by Daniel Baugh in "The Politics of British Naval Failure, 1775–1778," *American Neptune* 52 (1992): 221–46; see also Syrett, *Royal Navy*, esp. 61–91.

10. For the prompt ordering of frigates in the two previous wars, see Daniel A. Baugh, *Naval Administration in the Age of Walpole* (Princeton, N.J.: Princeton University Press, 1966), appendix 2; *Naval Administration, 1715–1750* (London: Navy Records Society, 1977), 193, 217–18; Richard Middleton, *The Bells of Victory: The Pitt-Newcastle Ministry and the Conduct of the Seven Years' War, 1757–1762* (Cambridge: Cambridge University Press, 1985), 108–9.

11. See G. R. Barnes and J. H. Owen, *The Private Papers of John, Earl of Sandwich* (London: Navy Records Society, 1932–38), 1: 20, 21, 23.

12. See Tracy, *Navies*, 38–41, 126–58.

13. See R. J. B. Knight, *Portsmouth Dockyard Papers, 1774–1783: The American War* (Portsmouth: Portsmouth Record Series, 1987), xliv–xlvi, 35, 45–50, 156–57. A shortage of skilled shipwright labor was the most constraining factor in British naval expansion for the whole of the century; also R. J. B. Knight, "The Building and Maintenance of the British Fleet during the Anglo-French Wars, 1688–1815," in Martine Acerra, José Merino, and Jean Meyer (eds.), *Les Marines de Guerre Européenes XVII–XVIII Siècles* (Paris: Presses de l'Université de Paris—Sorbonne, 1985), 38–39.

14. See H. M. Scott, "The Importance of Bourbon Naval Reconstruction to the Strategy of Choiseul after the Seven Years War," *International History Review* 1 (1979): 17–35; Tracy, *Navies*, 69–99; Jan Glete, *Navies and Nations: Warships, Navies and State Building in Europe and America, 1500–1860* (Stockholm: Almqvist and Wiksell International, 1993), 1: 285–86.

15. P. L. C. Webb, "The Rebuilding and Repair of the Fleet, 1783–1793," *Bulletin of the Institute of Historical Research* 50 (1977): 201–2.

16. From the evidence in these tables, it is clear that I take issue with Nicholas Tracy when he states (34–35) that Sandwich "more than held his own in the repair and expansion of ships."

17. See Knight, "Building and Maintenance," 42–43.

18. The phrase was used by William B. Wilcox in "Arbuthnot, Gambler and Graves: 'Old women' of the Navy," in G. Billias (ed.), *George Washington's Opponents* (New York: Morrow, 1969), 262.

19. The following arguments are based on J. H. Broomfield, "Lord Sandwich at the Admiralty Board: Politics and the British Navy, 1771–1778," in *Mariner's Mirror* 51 (1965): 7–17; and John A. Davies, "An Inquiry into Faction among British Naval Officers during the War of the American Revolution" (M.A. thesis, University of Liverpool, 1964).

20. See Kenneth Breen, "Divided Command: The West Indies and North America, 1780–1781," in Black and Woodfine (eds.), *British Navy*, 191–206.

21. N. A. M. Rodger, *Insatiable Earl*, 191; also Rodger, *The Wooden World: An Anatomy of the Georgian Navy* (London: Collins, 1986), 302.

22. See I. R. Christie, *The End of North's Ministry, 1780–1782* (London: MacMillan, 1958), 176–77; Davies, "Enquiry into Faction," 3–13.

23. Hilariously so, on at least one occasion. The account survives of a cabinet

meeting in 1780, after a good dinner, at which the decision to recall the Dutch ambassador was made; North was asleep throughout the meeting, though he was little better than the others. See H. M. Scott, "Sir Joseph Yorke: The Politics and Origins of the Fourth Dutch War," *Historical Journal* 31 (1988): 571–72.

24. Baugh, "British Naval Failure," 240; Mackesy, *War for America*, 283–84; Rodger, *Insatiable Earl*, 224–31, demonstrates the complexities of his political position.

25. Knight, "Building and Maintenance," 41–44. The Navy Board hardly ever used the northeast of Britain, yet this area was the most dynamic shipbuilding area in England.

26. Vergennes quoted in Dull, *French Navy*, 316–17. His chapter headings illustrate the sudden reversal of French fortunes and their loss of morale and political will: Chapter 8: "1781—the *Annus Mirabilis*"; Chapter 9: "1782—Disintegration and Reprieve." Nevertheless, a comparison with the tables in this paper and those in Dull (appendix C, pp. 352–55, and appendix D, pp. 356–58) show that the French were keeping up with construction and repairs up to 1782; but after that year they knew they could no longer compete. (I am grateful to Michael Duffy for some refinements on this point.) This differs from the interpretation of Jose P. Merino Navarro, *La Armada Espagnola en el Siglo XVIII* (Madrid: Fundacion Universitaria Espagnola, 1981), 357–58; his graphs demonstrate that Spanish construction declined relative to Britain as early as 1775, and French construction from 1780.

27. Dull, *French Navy*, 176, 257, 291.

28. J. E. Talbott, "Copper, Salt and the Worm," *Navy History* 3 (1989): 53; see also R. J. B. Knight, "The Introduction of Copper Sheathing into the Royal Navy, 1779–1786," in *Mariner's Mirror* 59 (1973): 299–309. French copper does seem to have been inferior; see Acerra and Meyer, *Marines*, 78–79.

29. Alan Jamieson, *War in the Leeward Islands, 1775–1783* (D.Phil., Oxford University, 1981), 47–50, 72, 245–46.

30. The most useful technical authority on the carronade is Brian Lavery, *The Army and Fitting of English Ships of War, 1600–1815* (London: Conway Maritime Press, 1987), 104–9, 123–25. For the problems of its introduction and acceptance, see John E. Talbott, "The Rise and Fall of the Carronade," *History Today* 39 (August 1989): 25–30.

31. Baugh, "Command of the Sea," 161.

32. John Brewer, *The Sinews of Power: War, Money and the English State, 1688–1783* (London: Unwin Hyman, 1989), 114–21, is especially illuminating on the state's dependence on credit and its ability to convert wartime short-term liabilities into long-term publicly funded credit. See also 175–78, 197–98.

33. William H. McNeill, *The Pursuit of Power* (Chicago: University of Chicago Press, 1982), 181.

34. Baugh, "Command of the Sea," 163.

35. John Sinclair, *Thoughts on the Naval Strength of the British Empire* (London: T. Cadell, 1782), 14–15.

Chapter 2

1. The literature on the Prussian Reform Era is very extensive. Most important are the documentary publications concerning the work of various government agencies and the memoirs, writings, and correspondence of such figures as Stein, Hardenberg, and Scharnhorst. Despite severe archival losses during the Second World War, new material continues to be published. Important examples are: Rudolf Ibbeken, *Preussen 1807–1813* (Cologne-Berlin: Grote, 1970); and the wide-ranging edition of writings by and documents concerning Clausewitz: Werner Hahlweg (ed.), *Carl von Clausewitz: Schriften—Aufsätze—Studien—Briefe* (Göttingen: Vandenhoeck and Ruprecht, 1966, 1990), 2 vols. The classic biographies of the leading figures continue to be important sources not only for their immediate subjects but also for the period as a whole. The most succinct account of the reform era in English is the translation of Friedrich Meinecke's famous essay in political, cultural, and intellectual history, Peter Paret (ed.), *The Age of German Liberation* (Berkeley: University of California Press, 1977). For the following analysis, I have also drawn on my two monographs, *Yorck and the Era of Prussian Reform* (Princeton, N.J.: Princeton University Press, 1966) and *Clausewitz and the State*, rev. ed. (Princeton, N.J.: Princeton University Press, 1985).

2. A convenient summary of the Jena campaign, with excellent maps, is given in Vincent J. Esposito and John Robert Elting (eds.), *A Military History and Atlas of the Napoleonic Wars* (New York: Praeger, 1964), unpaginated, maps 57–68.

3. For the work of the commission, see Rudolf Vaupel (ed.), *Das Preussische Heer vom Tilsiter Frieden bis zur Befreiung: 1807–1814* (Leipzig: S. Hirzel, 1938). Part 2 of *Die Reorganisation des Preussischen Staates unter Stein und Hardenberg*.

4. On the centennial of the defeats, the historical section of the Great General Staff produced a detailed documentary account of the investigations: *1806: Das Preussische Offizierkorps und die Untersuchung der Kriegsereignisse* (Berlin: Mittler und Sohn, 1906).

5. For a discussion of these nonexempt groups, see my essay "Conscription and the End of the Old Regime in France and Prussia," in Wilhelm Treue (ed.), *Geschichte als Aufgabe: Festschrift für Otto Büsch* (Berlin: Colloquium Verlag, 1988), now also in Peter Paret, *Understanding War: Essays on Clausewitz and the History of Military Power* (Princeton, N.J.: Princeton University Press, 1992).

6. One of the most interesting contemporary analyses of this significant change is Clausewitz's essay "Uber die politischen Vortheile und Nachtheile der Preussischen Landwehr," written in the vain hope of reversing the process. An English version of this essay is included in Peter Paret and Daniel Moran (eds.), *Carl von Clausewitz, Historical and Political Writings* (Princeton, N.J.: Princeton University Press, 1992).

Chapter 3

1. The best history of the Boer War is Thomas Pakenham's, *The Boer War* (London: Weidenfeld and Nicolson, 1979). See also Edgar Holt, *The Boer War* (London: Putnam, 1958); Rayne Kruger, *Good-bye Dolly Gray: The Story of the*

Boer War (London: Cassell, 1959); Byron Farwell, *The Great Anglo-Boer War* (New York: Harper and Row, 1976).

2. George Wyndham to his mother, 5 October 1899. J. W. Mackail and Guy Windham, *Life and Letters of George Wyndham* (London: Hutchinson, 1926), 1: 361.

3. Wolseley to Lady Wolseley, 29 September 1899. Quoted in Edward M. Spiers, *The Army and Society, 1815–1914* (London: Longman, 1980), 237.

4. Quoted in Franlyn Arthur Johnson, *Defence by Committee: The British Committee of Imperial Defence, 1885–1959* (London: Oxford University Press, 1960), 44.

5. Sir William Robertson, *From Private to Field-Marshal* (London: Constable, 1921), 97.

6. Pakenham, *Boer War*, 572; Thomas G. Fergusson, *British Military Intelligence, 1870–1914: The Development of a Modern Intelligence Organization* (London: Arms and Armour, 1984), 114.

7. Quoted in Brian Bond, *The Victorian Army and the Staff College, 1854–1914* (London: Eyre Methuen, 1972), 181.

8. The committees in question were: the Coloniai Defense Committee (created in 1885), the Joint Naval and Military Committee (1891), and the Standing Defence Committee of the Cabinet (1895). The best introduction to them and their successors is still John Ehrman, *Cabinet Government and War, 1890–1940* (Cambridge: Cambridge University Press, 1958).

9. Cd. 1790 (1903), *Minutes of Evidence Taken before the Royal Commission on the War in South Africa*, Vol. 1, Qs. 9035–36, 9041.

10. Middleton Papers. Roberts to St. John Broderick, 26 November 1900. P.R.O. 30/67/6.

11. W. S. Hamer, *The British Army: Civil-Military Relations, 1885–1905* (Oxford: Clarendon Press, 1970), 187–89.

12. Ehrman, *Cabinet Government and War*, 24.

13. Sanders Mss. Balfour to St. John Broderick, 31 October 1900. Bodleian Mss. Eng. Hist. C. 732.

14. See Johnson, *Defence by Committee*, 52–55.

15. See Spiers, *Army and Society*, 243–44, 250–51.

16. Sanders Mss. Acland Hood to Sanders, 2 September 1903. Bodleian Mss. Eng. Hist. C. 741.

17. Cd. 1790, *Minutes of Evidence*, Qs. 1827, 2444.

18. Cd. 1790, *Minutes of Evidence*, Qs. 4494–95, 4504, 47.

19. Cd. 1790, *Minutes of Evidence*, Qs. 6097–98, 6099.

20. Sanders Mss. Selborne to Balfour, 27 September 1903. Bodleian Mss. Eng. Hist. C. 742.

21. See James Lees-Milne, *The Enigmatic Edwardian: The Life of Reginald 2nd Viscount Esher* (London: Sidgwick and Jackson, 1986), 144–46.

22. Sanders Mss. Balfour to the duke of Devonshire, 24 September 1903. Bodleian Mss. Eng. Hist. C. 743.

23. For the work of the W.O.(R.)C., see John Gooch, *The Plans of War: The General Staff and British Military Strategy c.1900–1916* (London: Routledge and Kegan Paul, 1974), 32–59.

24. Cd. 1790, *Minutes of Evidence*, Q. 9195.

25. Cd. 1790, *Minutes of Evidence*, Q. 6842.

26. Cd. 1790, *Minutes of Evidence*, Qs. 943, 982, 1070–72, 4399 [on mobilization procedures]; 2825–26, 2856–57, 2861, 9522–34 [on Admiralty/War Office cooperation].

27. Fergusson, *British Military Intelligence*, 113–15.

28. Cd. 1790, *Minutes of Evidence*, Qs. 5126–27.

29. Cd. 1790, *Minutes of Evidence*, Q. 5163.

30. H. H. R. Bailes, "The Influence of Continental Examples and Colonial Warfare upon the Reform of the Late-Victorian Army" (Ph.D. diss., University of London, 1980), ch. 8.

31. Shelford Bidwell and Dominic Graham, *Fire-Power: British Army Weapons and Theories of War, 1904–1945* (London: Allen and Unwin, 1982), 2.

32. Cd. 1791 (1903), *Minutes of Evidence Taken before the Royal Commission on the War in South Africa*, Vol. 2, Q. 1599.

33. David French, *British Economic and Strategic Planning, 1905–1915* (London: Allen and Unwin, 1982), 39–47.

34. Quoted by Pakenham, *Boer War*, 195.

35. G. F. R. Henderson, *The Science of War* (London: Longmans, Green, 1906), 351, 372; Cd. 1792 (1903), *Minutes of Evidence Taken before the Royal Commission on the War in South Africa*, Vol. 3, Q. 13 941.

36. Bailes, "Reform of the Late-Victorian Army," ch. 4.

37. See Spiers, *Army and Society*, 245, 247 [on Roberts]. It is interesting to compare Hamilton's evidence to the Elgin Commission [Cd. 1791, esp. Q. 13 941] with his book *The Fighting of the Future* (London: Kegan Paul, Trench, 1895).

38. Sir G. F. Ellison, "Lord Roberts and the General Staff," *Nineteenth Century and After* 112 (December 1932): 729; Spiers, *Army and Society*, 247.

39. Cd. 1791, *Minutes of Evidence*, Q. 13 941.

40. For a concise summary of the *arme blanche* controversy, see Brian Bond, "Doctrine and Training in the British Cavalry, 1870–1914" in Michael Howard (ed.), *The Theory and Practice of War* (London: Cassell, 1965), 95–125.

41. Stephen Badsy, "Fire and the Sword: The British Army and the Arme Blanche Controversy, 1871–1921" (Ph.D. diss., University of Cambridge, 1982), 139.

42. Badsy, "Fire and the Sword," 146.

43. Esher Papers. Haig to Esher, 21 March 1903. Army Letters, Vol. 1.

44. Cd. 1791, *Minutes of Evidence*, Q. 16 924.

45. See R. J. Q. Adams and Philip P. Poirier, *The Conscription Controversy in Britain, 1900–18* (London: Macmillan, 1987).

46. On this issue, see John Gooch, *The Prospect of War: Studies in British Defence Policy, 1847–1942* (London: Cass, 1981), 92–115; Edward M. Spiers, *Haldane: An Army Reformer* (Edinburgh: Edinburgh University Press, 1980); Ian F. W. Beckett, *Riflemen Form: A Study of the Rifle Volunteer Movement, 1859–1908* (Aldershot: Ogilby Trusts, 1982).

47. Quoted in Bond, "Doctrine and Training," 188–89.

48. Bond, "Doctrine and Training," 197.

49. See Gooch, *Plans of War*, 32–164.

50. Quoted in Ian Worthington, "Antecedent Education and Officer Recruitment: An Analysis of the Public School–Army Nexus, 1849–1908" (Ph.D. diss., University of Lancaster, 1982), 325.

51. Esher Papers. Haig to Esher, 11 January 1906. Army Letters, Vol. 3.

52. Cd. 1790, *Minutes of Evidence*, Qs. 9381–82.

53. Worthington, "Antecedent Education," 242–65, 276, 282.

54. Robertson, *From Private to Field-Marshal*, 95; Bond, *Victorian Army and the Staff College*, 187; E. K. G. Sixsmith, *British Generalship in the Twentieth Century* (London: Arms and Armour Press, 1970), 13.

55. N. d'Ombrain, *War Machinery and High Policy: Defence Administration in Peacetime Britain, 1902–1914* (London: Oxford University Press, 1973), 141–43 [quotation from 143]; Tim Travers, "The Hidden Army: Structural Problems in the British Officer Corps, 1900–1918," *Journal of Contemporary History* 17 (July 1982): 523–44 [quotation from 537].

56. Bailes, "Reform of the Late-Victorian Army," 76.

57. Bidwell and Graham, *Fire-Power*, 52. For an excellent survey of this question, see Michael Howard, "Men against Fire: The Doctrine of the Offensive in 1914" in Peter Paret (ed.), *Makers of Modern Strategy from Machiavelli to the Nuclear Age* (Princeton, N.J.: Princeton University Press, 1986), 510–26.

58. Sir Edward Bruce Hamley, *The Operations of War Explained and Illustrated* (Edinburgh: Blackwood, 1914), 406. The revision was done by Brigadier General L. E. Kiggell, and the new edition was reprinted in 1909 and 1914.

Chapter 4

1. Carlos Baker, *Ernest Hemingway: A Life Story* (New York: Scribner, 1969), 31–44, 53–54, 58–59, 66. Hemingway arrived in Italy in early June 1918. He saw no significant combat, taking no part in either the battles of the Piave or Vittorio Veneto.

2. Ronald Seth, *Caporetto: The Scapegoat Battle* (London: Macdonald, 1965), 13–19.

For a graphic depiction of these various military disasters, see Thomas E. Griess (ed.), *Campaign Atlas to the Great War*, West Point Military History Series (Garden City Park, N.Y.: Avery, 1986).

3. Lucio Ceva, *Le forze armate* (Turin: UTET, 1981), 152.

4. This was the conclusion of the official parliamentary commission of inquiry into the causes of the defeat. For the complete report, see *Relazione della Commissione d'inchiesta sugli avvenimenti che hanno determinato il ripiegamento del nostro esercito sul Piave: Dall'Isonzo al Piave, 24 ottobre–9 novembre 1917*, 3 vols. (Rome, 1919). For excerpts, see Shepard B. Clough and Salvatore Saladino, *A History of Modern Italy: Documents, Readings and Commentary* (New York: Columbia University Press, 1968), 337–42.

5. Franco Molfese, *Storia del brigantaggio dopo l'Unità* (Milan: Feltrinelli, 1964); A. de Jaco, *Il brigantaggio meridionale* (Rome: Riuniti, 1969); John Whittam, *The Politics of the Italian Army, 1861–1918* (London: Croon Helm, 1977), 73–86.

6. For an overview of the political, economic, and social aspects of the Risorgimento, see Harry Hearder, *Italy in the Age of the Risorgimento, 1790–1870* (London: Longman, 1983); Stuart Woolf, *A History of Italy, 1700–1860: The Social Constraints of Political Change* (London: Methuen, 1986); Denis Mack Smith, *Cavour* (New York: Knopf, 1985). The military history of the Risorgimento is best told in Piero Pieri, *Storia militare del Risorgimento: Guerre e Insurrezioni* (Turin: Einaudi, 1962).

7. For the history of Italy during its painful first decades of unification, see Martin Clark, *Modern Italy, 1871–1982* (London: Longman, 1984), 1–177; Christopher Seton-Watson, *Italy from Liberalism to Fascism, 1870–1925* (London: Methuen, 1967), 3–410.

For studies of the Italian army between the Risorgimento and the First World War, see John Gooch, *Army, State and Society in Italy, 1870–1915* (London: Macmillan, 1989); Massimo Mazzetti, *L'esercito italiano nella triplice alleanza* (Naples: Edizioni Scientifiche Italiane, 1974); Giorgio Rochat and Giulio Massobrio, *Breve storia dell'esercito italiano dal 1861 al 1943* (Turin: Einaudi, 1978), 5–168; Ceva, *Le forze armate*, 41–113; Whittam, *Politics of the Italian Army*, 11–188.

8. Denis Mack Smith, *Italy and Its Monarchy* (New Haven: Yale University Press, 1989), 201–14; Ceva, *Le forze armate*, 154–62; Seton-Watson, *Italy from Liberalism to Fascism*, 413–50; Clough and Saladino, *History of Modern Italy*, 66–70, 303–26; Piero Melograni, *Storia politica della grande guerra, 1915–1918* (Bari: Laterza, 1969), 3–11.

9. Gooch, *Army, State and Society*, 156–70; Giorgio Rochat, "La preparazione dell'esercito italiano nell'inverno 1914–1915 in relazione alle informazioni disponsibili sulla guerra di posizione," *Il Risorgimento* 13 (February 1961): 10–32; Pierluigi Scole, "Le lezioni tattiche del fronte occidentale e il loro mancato riflesso sulla guerra italiana (1915–1917)" (Ph.D. diss., University of Pavia, 1990), 37–39, 49–55.

10. Ceva, *Le forze armate*, 112–20; Gooch, *Army, State and Society*, 147–64; Angelo Gatti, *Un italiano a Versailles (Dicembre 1917–Febbraio 1918)* (Milan: Ceschina, 1958), 356; Lucio Ceva and Andrea Curami, *La meccanizzazione dell'esercito fino al 1943*, 2 vols. (Rome: Stato Maggiore Esercito, Ufficio Storico, 1989), 1: 34–35; Piero Pieri, *L'Italia nella prima guerra mondiale* (Turin: Einaudi, 1968), 89.

11. Giorgio Rochat, *Gli arditi della grande guerra; Origini, battaglie e miti* (Gorizia: Editrice Goriziana, 1990), 36; John Gooch, "Italy during the First World War," in Allan R. Millett and Williamson Murray (eds.), *Military Effectiveness*, 3 vols. (Boston: Allen and Unwin, 1988), 1, *The First World War*, 170–72, 177–79.

12. James Edmonds and H. R. Davies, *Military Operations, Italy 1915–1919* (London: Her Majesty's Stationery Office, 1949), 11; Ceva, *Le forze armate*, 123.

13. Gooch, "Italy during the First World War," 170–71; Gooch, *Army, State and Society*, 157, 170; Whittam, *Politics of the Italian Army*, 195; Gatti, *Un italiano a Versailles*, 333, 419–20, 428.

14. Melograni, *Storia politica della grande guerra*, 120–24, 243–62, 283–300; Ceva, *Le forze armate*, 164–65; Whittam, *Politics of the Italian Army*, 196–98.

15. For events on the Italian Front and the politics behind it between May

1915 and October 1917, see Pieri, *L'Italia nella prima guerra mondiale*, 80–152; Emilio Faldella, *La grande guerra*, 2 vols. (Milan: Longanesi, 1978), 1, *Le battaglie dell'Isonzo (1915–1917)*; Ceva, *Le forze armate*, 122–36; Seth, *Caporetto*, 52–135; Seton-Watson, *Italy from Liberalism to Fascism*, 450–77.

16. Istvan Deak, *Beyond Nationalism: A Social and Political History of the Habsburg Officer Corps, 1848–1918* (New York: Oxford University Press, 1990), 190–98; Gunther E. Rothenberg, *The Army of Francis Joseph* (West Lafayette, Ind.: Purdue University Press, 1976), 180–206.

17. Seth, *Caporetto*, 52–61; Virgilio Ilari, *Storia del servizio militare in Italia*, 3 vols. (Rome: CEMISS, 1990), 2, *La "nazione armata" (1871–1918)*, 460–79; Ceva, *Le forze armate*, 162–73.

18. Rothenberg, *Army of Francis Joseph*, 205–6; Cyril Falls, *The Battle of Caporetto* (Philadelphia: Lippincott, 1966), 17–18; Seth, *Caporetto*, 117–18, 137.

19. Melograni, *Storia politica della grande guerra*, 374, 386–88; Faldella, *La grande guerra*, 2, *Da Caporetto al Piave (1917–1918)*, 316–27; Whittam, *Politics of the Italian Army*, 201–2.

20. Melograni, *Storia politica della grande guerra*, 329–42; Gianni Toniolo, *An Economic History of Liberal Italy, 1850–1918* (London and New York: Routledge, 1990), 131; Seth, *Caporetto*, 130–31.

21. Seton-Watson, *Italy from Liberalism to Fascism*, 468–77; Ceva, *Le forze armate*, 127–30; Faldella, *La grande guerra*, 2: 298–316; Melograni, *Storia politica della grande guerra*, 389–93; Gatti, *Un italiano a Versailles*, 52.

22. Odoardo Marchetti, *Il servizio informazione dell'esercito italiano nella grande guerra* (Rome: Tipgrafia regionale, 1937), 181; Bruce I. Gudmundsson, *Stormtroop Tactics: Innovation in the German Army, 1914–1918* (New York: Praeger, 1989), 125–38, 152 n. 4; Melograni, *Storia politica della grande guerra*, 393–417; Falls, *The Battle of Caporetto*, 24–44; Emilio Faldella, *La grande guerra*, 2: 21–203; Piero Pieri and Giorgio Rochat, *Pietro Badoglio* (Turin: UTET, 1974), 241–370; Ceva, *Le forze armate*, 136–40.

23. Faldella, *La grande guerra*, 2: 204–47; Pieri and Rochat, *Pietro Badoglio*, 370–400; Melograni, *Storia politica della grande guerra*, 423–54; Mario Bernardi, *Di qua e di là dal Piave: Da Caporetto a Vittorio Veneto* (Milan: Mursia, 1989), 13–34. For detailed descriptions of the battle from the Italian side, see Angelo Gatti, *Caporetto* (Bologna: Il Mulino, 1964) and Emilio Faldella, *Caporetto: Le vere cause di una tragedia* (Bologna: Cappelli, 1967).

24. Ilari, *Storia del servizio militare in Italia*, 2: 432; Ceva, *Le forze armate*, 143; Faldella, *La grande guerra*, 2: 292–98. Recent research by Giuliana Procacci, mentioned in a letter from Lucio Ceva to the author, suggests that Italian prisoners captured during the Caporetto offensive may have numbered over 300,000.

25. Ministero degli Affari Esteri, Commissione per la pubblicazione dei documenti diplomatici, *I documenti diplomatici Italiani*, Fifth Series [hereafter DDI, followed by volume and document number], vol. 9, no. 299; Olindo Malagodi, *Conversazioni della guerra*, 2 vols. (Milan: Ricciardi, 1960), 1: 207; Luigi Cadorna, *Pagine polemiche* (Milan: Garzanti, 1951), 253–56; Gatti, *Un italiano a Versailles*, 313.

26. Faldella, *La grande guerra*, 2: 247–77; Seth, *Caporetto*, 162–80; DDI, vol. 9,

no. 359; Seton-Watson, *Italy from Liberalism to Fascism*, 491; Ceva, *Le forze armate*, 143; *Les Armées françaises dans la grande guerre*, tome 6, vol. 1 (Paris, 1932), 93–94; Edmonds and Davies, *Military Operations*, 88–98.

27. DDI, vol. 9, no. 391; Edmonds and Davies, *Military Operations*, 73–81; Mack Smith, *Italy and Its Monarchy*, 229–30; Melograni, *Storia politica della grande guerra*, 454–58; Gatti, *Un italiano a Versailles*, 22, 31–33; Malagodi, *Conversazioni della guerra*, 1: 206.

28. Faldella, *La grande guerra*, 2: 282–88, 333–36; Edmonds and Davies, *Military Operations*, 99–116, 142; Norman Gladden, *Across the Piave: A Personal Account of the British Forces in Italy, 1917–1919* (London: Her Majesty's Stationery Office, 1971), 28–30; *Les Armées françaises*, tome 6, vol. 1, pp. 112–13. For German operations in Italy in the fall of 1917, see Oberkommando des Heeres, *Der Weltkrieg* (Berlin, 1942), 13: 218–308. For Austro-Hungarian operations, see *Österreich-Ungarns letzter Krieg*, 6, *Das Kriegsjahr 1917* (Vienna, 1936).

29. Lucio Ceva, "La battaglia dei '3 monti,' " *Storia e Dossier* 3 (January 1989): 35–39.

30. Melograni, *Storia politica della grande guerra*, 305, 445–48; Ceva, *Le forze armate*, 167; Ilari, *Storia del servizio militare in Italia*, 2: 432–33.

31. DDI, vol. 9, nos. 438, 625; ibid., vol. 10, no. 233; Melograni, *Storia politica della grande guerra*, 463–64.

32. Pieri and Rochat, *Pietro Badoglio*, 414–18, 428; Gatti, *Caporetto*, 379–81, 388–90, 453–54, 458; Gatti, *Un italiano a Versailles*, 26–27, 210–11.

33. Melograni, *Storia politica della grande guerra*, 469–75; Seton-Watson, *Italy from Liberalism to Fascism*, 482–85; Whittam, *Politics of the Italian Army*, 203–4.

34. Seton-Watson, *Italy from Liberalism to Fascism*, 486–89, 497; Edmonds and Davies, *Military Operations*, 119–20; John Terraine, *The U-Boat Wars, 1916–1945* (New York: Putnam, 1990), 98–99, 130–32.

35. Faldella, *La grande guerra*, 2: 339; Ceva, *Le forze armate*, 145–46; Rochat, *Gli arditi della grande guerra*, 60–64; Vincenzo Gallinari, *L'esercito italiano nel primo dopoguerra, 1918–1920* (Rome: Stato Maggiore Esercito, Ufficio Storico, 1980), 223–24.

36. Ceva, *Le forze armate*, 144–45; Ilari, *Storia del servizio militare in Italia*, 2: 434; Rochat, *Gli arditi della grande guerra*, 64–65.

37. Melograni, *Storia politica della grande guerra*, 501–21; Whittam, *Politics of the Italian Army*, 205–6.

38. DDI, vol. 10, nos. 419, 510; Falls, *Battle of Caporetto*, 151–55.

39. Ceva, *Le forze armate*, 146–49; Lucio Ceva, "La battaglia del Solstizio," *Storia e Dossier* 3 (June 1988); Faldella, *La grande guerra*, 2: 352–66; Rothenberg, *Army of Francis Joseph*, 213; Falls, *Battle of Caporetto*, 157–73; Deak, *Beyond Nationalism*, 202; Gladden, *Across the Piave*, 131–33; Bernardi, *Di qua e di là dal Piave*, 125–51. For Austro-Hungarian operations in 1918, see *Österreich-Ungarns letzter Krieg*, 7, *Das Kriegsjahr 1918* (Vienna, 1937).

40. Arthur J. May, *The Passing of the Hapsburg Monarchy, 1914–1918*, 2 vols. (Philadelphia: University of Pennsylvania Press, 1966), 2: 732–64; C. A. Macartney, *The Habsburg Empire* (New York: Macmillan, 1969), 830.

41. Seton-Watson, *Italy from Liberalism to Fascism*, 500–501; Sidney Sonnino,

Diario 1916–1922 (Bari: Laterza, 1972), 301–5; Pieri and Rochat, *Pietro Badoglio,* 440–42; Pieri, *L'Italia nella prima guerra mondiale,* 197–98; *Les Armées françaises dans la grande guerre,* tome 6, vol. 2 (Paris, 1934), 362–66; ibid., tome 7, vol. 2 (Paris, 1938), 356–58; Edmonds and Davies, *Military Operations,* 242–49; Ceva, *Le forze armate,* 149–51; Lucio Ceva, "La grande guerra nel Veneto: Scrittori e memorialisti," *La Cultura* 1 (1988): 122–24. Parri would go on to become a major leader of the opposition to Mussolini's dictatorship and found the anti-Fascist "Justice and Liberty" movement and the Action Party. In 1943–45, Parri served as the de facto commander of the Italian partisans. In June–December 1945, Parri held office as Italy's first postwar prime minister.

42. Ceva, *Le forze armate,* 151–52, 166 n. 1; Pieri and Rochat, *Pietro Badoglio,* 442–49; Pieri, *L'Italia nella prima guerra mondiale,* 198–204; Z. A. B. Zeman, *The Break-Up of the Habsburg Empire* (London: Oxford University Press, 1961), 217–45; Rothenberg, *Army of Francis Joseph,* 216–18; Deak, *Beyond Nationalism,* 202–3; Gladden, *Across the Piave,* 164–71; Gallinari, *L'esercito italiano nel primo dopoguerra,* 10–17; Bernardi, *Di qua e di là dal Piave,* 170–83.

43. Sonnino, *Diario 1916–1922,* 308–13, 316–20, 326, 331–32; Seton-Watson, *Italy from Liberalism to Fascism,* 504–10, 527–36.

44. Seton-Watson, *Italy from Liberalism to Fascism,* 510–27, 536–60.

45. Literature in Italian on the rise of fascism is voluminous. The best works include: Luigi Salvatorelli and Giovanni Mira, *Storia d'Italia nel periodo fascista* (Turin: Einaudi, 1957), 15–221; Enzo Santarelli, *Storia del fascismo,* 3 vols. (Rome: Riuniti, 1973), 1: 85–321; Renzo De Felice, *Mussolini il rivoluzionario* (Turin, 1965); Renzo De Felice, *Mussolini il fascista,* vol. 1, *La conquista del potere, 1921–1925* (Turin: Einaudi, 1966), 3–387; Emilio Gentile, *Storia del Partito Fascista, 1919–1922: Movimento e milizia* (Rome-Bari: Laterza, 1989). In English, see Adrian Lyttelton, *The Seizure of Power: Fascism in Italy, 1919–1929* (Princeton, N.J.: Princeton University Press, 1987), 1–93; Seton-Watson, *Italy from Liberalism to Fascism,* 570–629; Clark, *Modern Italy,* 203–21; Denis Mack Smith, *Mussolini* (New York: Knopf, 1983), 32–51; Philip V. Cannistraro and Brian R. Sullivan, *Il Duce's Other Woman* (New York: William Morrow, 1993), 196–264.

Chapter 5

1. This wider theme will be more systematically developed in a forthcoming volume of essays on the politics of the French officer corps, Philip C. F. Bankwitz (ed.), *Conflicts of Duty: Civil-Military Relations in Twentieth Century France* (forthcoming).

2. See Jean Lacouture, *De Gaulle,* vol. 1, *The Rebel (1890–1944)* (London and New York: Collins-Harvill, 1990); Arthur Clendenin Robertson, *La Doctrine du général de Gaulle* (Paris: Arthème Fayard, 1959).

3. A point well drawn out in Maurice Vaïsse, *1961: Alger, le putsch,* Coll. "La mémoire du siècle" (Brussels: Editions Complex, 1983), 91–92.

4. See Philip C. F. Bankwitz, *Maxime Weygand and Civil-Military Relations in Modern France* (Cambridge, Mass.: Harvard University Press, 1967), 305–14.

5. A point emphasized in Orville D. Menard, *The Army and the Fifth Republic* (Lincoln: University of Nebraska Press, 1967), esp. 209. On Operation Resurrection, see Philip M. Williams, *Wars, Plots and Scandals in Postwar France* (Cambridge: Cambridge University Press, 1970), 129–66; Jean Ferniot, *De Gaulle et le 13 mai* (Paris: Plon, 1965), 366–486; François Mitterrand, *Le Coup d'état permanent* (Paris: Plon, 1964), 273–79. The confessions of the two key civilian and military plotters, Michel Debré and General Roger Miquel, are found, respectively, in Odile Rudelle, *Mai 58: De Gaulle et la république* (Paris: Plon, 1988), 254–55, and Miquel, *Opération Resurrection* (Paris: Editions France-Empire, 1975), esp. 126–27.

6. On the visit by de Gaulle and his entire family to French military headquarters at Baden-Baden on 29 May 1968 to consult the commander of French forces in Germany on the loyalty of these forces to de Gaulle, see the damning memoir of their commander, a veteran of Indochina and Algeria, General Jacques Massu, *Baden, mai 1968: Souvenirs d'une fidelité gaulliste* (Paris: Plon, 1983); also Georges Menant, "Il y a dix ans: Le jour où de Gaulle a disparu, 29 mai 1968," *Paris-Match*, no. 1510 (5 mai 1978): 90–93, 110.

7. See Jacques Branet, *L'Escadron: Carnets d'un cavalier* (Paris: Flammarion, 1968); Suzanne Massu, *Quand j'étais Rochambelle: De New York à Berchtesgaden* (Paris: Grasset, 1969); and Erwan Bergot, *La 2e DB*, Coll. "Troupes de Choc" (Paris: Presses de la Cité, 1977); Maréchal Jean de Lattre de Tassigny, *Histoire de la Première Armée française: Rhin et Danube* (Paris: Plon, 1949); Major-General Sir Guy Salisbury-Jones, *So Full a Glory: A Biography of Marshal de Lattre de Tassigny* (New York: Frederick A. Praeger, 1955), 114–206.

8. An excellent overview of the French war in Indochina is provided by Bernard Fall, *Street without Joy: Indochina at War, 1945–54* (Harrisburg, Penn.: Stackpole, 1961).

9. See General Vo Nguyen Giap, *People's Army, People's War: The Viet Cong Insurrection Manual for Underdeveloped Countries*, foreword by Roger Hilsman (New York: Frederick A. Praeger, 1962); Ho Chi Minh, *Action et révolution, 1920–1967* (Paris: Union Générale d'Editions, 1968).

10. Quoted in Jacques Dalloz, *La Guerre d'Indochine* (Paris: Seuil, 1987), 106.

11. The persistent rumor of communist responsibility for the crash of Leclerc's airplane arose out of the Second Armored Division's mission to Chateauroux in March 1945, allegedly to counter communist-inspired social agitation. This assignment effectively kept the unit from participating in the invasion of Germany until the last minute in May 1945—too late for the reward the Division felt it deserved. Communist resentment of the Second Armored Division's role in the liberation of Paris in August 1944 also plays a part in the question. See Jean-Julien Fonde, *Les Loups de Leclerc: Récit* (Paris: Plon, 1982), 291–303, 351–52.

12. Salisbury-Jones, *So Full a Glory*, 270–77.

13. See especially the notebooks of the most remarkable soldier of this entire period, Hélie Denoix de Saint-Marc, a captain in the Foreign Legion who was rescued in May 1945 near death from Buchenwald, where he had been imprisoned as a résistant at the age of nineteen. They are found in the biography by Lau-

rent Beccaria, *Hélie de Saint-Marc* (Paris: Perrin, 1988), 84–108. See also Pierre Sergent, *Paras-Légion: Le 2e BEP en Indochine* (Paris: Presses de la Cité, 1982), 53–90.

14. Quoted in General Raoul Salan, *Indochine rouge: Le Message de Ho Chi Minh* (Paris: Presses de la Cité, 1975), 31.

15. Salisbury-Jones, *So Full a Glory*, 241–65, 273–74.

16. Salan, *Indochine rouge*, 32. See also Alain Gandy, *Salan* (Paris: Perrin, 1990), 129–94.

17. Salisbury-Jones, *So Full a Glory*, 274.

18. For a discussion that compares French methods employed against guerrilla and insurgency warfare in Indochina and Algeria, see Peter Paret, *French Revolutionary Warfare from Indochina to Algeria* (London: Pall Mall Press, 1964).

19. Salan, *Indochine rouge*, 32; see also Gandy, *Salan*, 195–206.

20. Salan, *Indochine rouge*, 36. For accounts of the French approaches to the U.S. secretary of state, John Foster Dulles, on the A-bomb questions, compare General Henri Navarre, *Le Temps des vérités* (Paris: Plon, 1979), 445–47; and Georges Bidault, *Resistance: The Political Autobiography of Georges Bidault* (New York: Praeger, 1967), 195–98.

21. See Douglas Porch, *The French Foreign Legion* (New York: Harper Collins, 1991), 531–33; Colonel Raymond Muelle, *Le Bataillon des réprouvés: Indochine, 1949–1950*, Coll. "Troupes de Choc" (Paris: Presses de la Cité, 1990).

22. See General Henri Navarre, *L'Agonie de l'Indochine* (Paris: Plon, 1956), 61–86; George Herring, "La conduite de la guerre d'Indochine," in Denise Artaud and Lawrence Kaplan (eds.), *Dien Bien Phu: L'Alliance atlantique et la défense du Sud-Est asiatique* (Lyon: La Manufacture, 1989), 87–95. The best account of the fighting at France's *camp retranché* of Dien Bien Phu remains Bernard Fall, *Hell in a Very Small Place* (New York: J. B. Lippincott, 1967).

23. Quotations cited in Vaïsse, *1961: Alger, le putsch*, 52–53; see also the account by Hélie Denoix de Saint-Marc of the utter demoralization occurring in the National Assembly (Beccaria, *Saint-Marc*, 106).

24. See Stanley Hoffman, "Paradoxes of the French political community," in Stanley Hoffman, Charles P. Kindleberger, Laurence Wylie, Jesse R. Pitts, Jean-Baptiste Duroselle, François Goguel (eds.), *In Search of France* (Cambridge, Mass: Harvard University Press, 1963), 1–117.

25. Herbert Tint, *French Foreign Policy since the Second World War* (New York: St. Martin's Press, 1972), 191. On the Geneva peace conference that liquidated the French involvement in Indochina, see Robert F. Randle, *Geneva 1954: The Settlement of the Indochinese War* (Princeton, N.J.: Princeton University Press, 1969).

26. See Tint, *French Foreign Policy*, 197.

27. A concise modern history of the French settlement and its interaction with native Algerian society may be found in Benjamin Stora, *Histoire de l'Algérie coloniale, 1830–1954* (Paris: Editions La Découverte, 1991). See also David C. Gordon, *North Africa's French Legacy, 1954–1962*, Harvard Middle Eastern Monograph Series (Cambridge, Mass: Harvard University Press, 1962), esp. 6–46; Williams, *Wars, Plots and Scandals*, 167–84.

28. Vaïsse, *1961: Alger, le putsch*, 54–55; see also the Massu memoirs in Jacques Massu and Henri Le Mire, *Vérité sur Suez, 1956* (Paris: Plon, 1978).

29. Vaïsse, *1961: Alger, le putsch*, 55 (quoting Challe's memoirs, *Notre Révolte* [Paris: Presses de la Cité, 1968]).

30. Tint, *French Foreign Policy*, 194.

31. Excellent modern overviews of the war that provide perspective on the varied and brutal character of the conflict include: Alistair Horne, *A Savage War of Peace: Algeria, 1954–1962* (New York: Viking Press, 1977); John Talbott, *The War without a Name: France in Algeria, 1954–1962* (New York: Alfred A. Knopf, 1980).

32. Tint, *French Foreign Policy*, 191.

33. See Porch, *Foreign Legion*, 580–85. Massu's own account appeared in his memoirs, *La Vraie Bataille d'Alger* (Paris: Plon, 1971). The ruthlessness of French methods for attacking the urban guerrilla threat was exposed in the hard-hitting film made by the Italian director Gilo Pontecorvo, *The Battle of Algiers*—a film that was banned from cinema screens in France upon its release in 1966.

34. See John Phillips, "Sons of France's Forgotten Algerians Demand to Live in Dignity," *The Times* (London), 13 August 1991, 8; Georges Fleury, *Les Harkis* (Paris: Bernard Grasset, 1974).

35. Tony Smith, *The French Stake in Algeria, 1945–1962* (Ithaca, N.Y.: Cornell University Press, 1978), 133.

36. Mollet's statement in the French Socialist party's newspaper, *Le Populaire*, 16 April 1956 (quoted in Smith, *French Stake in Algeria*, 133).

37. Vaïsse, *1961: Alger, le putsch*, 54.

38. Smith, *French Stake in Algeria*, 133. For the conscript experience, see Erwan Bergot, *La Guerre des appelés en Algérie, 1956–1962* (Paris: Presses de la Cité, 1980).

39. General E. Jouhaud, *ô mon pays perdu* (Paris: Arthème Fayard, 1969); Jouhaud, *Serons-nous enfin compris?* (Paris: Arthème Fayard, 1974); a third volume of memoirs by Jouhaud was entitled *Ce que je n'ai pas dit* (Paris: Arthème Fayard, 1977).

40. Salan has written extensively on his part in both the Indochina War and the Algerian uprising, under a general title that emphasizes his preoccupation with the question of the collapse of the French Empire. See General Raoul Salan, *Mémoires: Fin d'un empire*, 4 vols. (Paris: Presses de la Cité, 1970–1974): 1, *Le Sens d'un engagement*; 2, *Le Vietminh mon adversaire, octobre 1946-octobre 1954*; 3, *Algérie française*; 4, *De Gaulle, Algérie et moi*. See also the April 1981 special issue of *Historia* devoted to Salan on the twentieth anniversary of his 1961 putsch attempt in Algiers against de Gaulle and the sympathetic modern biography by Gandy (*Salan*, esp. 31–54, 233–326).

41. See General A. Zeller, *Dialogues avec un général* (Paris: Plon, 1974).

42. The most celebrated case of torture is presented in Henri Alleg, *La Question* (Paris: Editions de Minuit, 1958). The most notable protests come from Jean-Jacques Servan-Schreiber, *Lieutenant en Algérie* (Paris: Plon, 1959) (Cf. his *Passions* [Paris: Fixot, 1991], 307–74); and from Jules Roy, *La Guerre d'Algérie* (Paris: Julliard, 1960); (Cf. his *J'accuse le général Massu* [Paris: Editions du Seuil, 1972] and his *Mémoires barbares* [Paris: Albin Michel, 1989], 551–608). General

discussion of torture is found in Porch, *Foreign Legion*, 585–89; and a recent overview that sets FLN atrocities against the unsavory record of the French paras in Olivier Drape, "Guerre d'Algérie: L'Histoire falsifiée," part 1, *Permanences*, no. 281, 25–32; part 2, *Permanences*, no. 282, 9–18.

43. See Smith, *French Stake in Algeria*, 133–37.

44. Jacques Soustelle, *L'Espérance trahie, 1958–1961* (Paris: Editions de L'Alma, 1962).

45. Tint, *French Foreign Policy*, 193.

46. See Rudelle, *Mai 58*, 161–213; Gandy, *Salan*, 281–303.

47. James H. Meisel, *The Fall of the Republic: Military Revolt in France* (Ann Arbor: University of Michigan Press, 1962), 200. Cf. Edgar S. Furniss, Jr., *De Gaulle and the French Army: A Crisis in Civil-Military Relations* (New York: Twentieth Century Fund, 1964); John Steward Ambler, *The French Army in Politics, 1945–1962* (Columbus: Ohio State University Press, 1966).

48. See Massu and Le Mire, *Vérité sur Suez;* Hélie Denoix de Saint-Marc in Beccaria, *Saint-Marc*, 184–85; Colonel Antoine Argoud, *La Décadence, l'imposture et la tragédie* (Paris: Arthème Fayard, 1974); Colonel Pierre Château-Joubert, *Feux et lumières sur ma trace* (Paris: Presses de la Cité, 1976); and the account of a ranker in both Indochina and Algeria, Sergeant Roland Barra, in Georges Fleury, *Donnez-moi la tourmente* (Paris: Bernard Grasset, 1987).

49. See Porch, *Foreign Legion*, 591, 607.

50. Maurice Mégret, "Fonction et intégration politiques de l'armée," in Centre de Sciences Politiques de l'Institut juridiques de Nice, *La Défense nationale* (Paris: Presses Universitaires, 1958), 157.

51. Pierre Sergent, *Je ne regrette rien* (Paris: Arthème Fayard, 1972). In the 1980s Sergent reemerged as a national political figure of notoriety, serving as a vice president of the executive committee of Jean-Marie Le Pen's far-right Front National party and securing election under the system of proportional representation to the French National Assembly from March 1986 to May 1988, as a Front National deputy for Orange. Author of numerous books on aspects of the history of the Foreign Legion, he has also turned his hand to military fiction, publishing *Les Voies de l'honneur* (Paris: Presses de la Cité, 1988).

52. Pierre Demaret and Christian Plume, *Target de Gaulle*, trans. Richard Barry (New York: Dial Press, 1975), 20.

53. See Jean Ferrandi, *600 Jours avec Salan et l'OAS* (Paris: Arthème Fayard, 1970); Gandy, *Salan*, 365–81.

54. Both Degueldre and Delhomme had served in Foreign Legion and special forces operations in Indochina before serving in Algeria and being recruited into the OAS.

55. The role of Susini, Ortiz, Delgueldre, and Delhomme in creating Delta is discussed in Rémi Kauffer, *OAS: Histoire d'une organisation secrète* (Paris: Arthème Fayard, 1986), 38–249; it also includes a biographical updating to 1986 of the OAS members and their opponents (387–401). The account by Susini is incomplete (*Histoire de l'OAS* [Paris: La Table Ronde, 1963], esp. 1–68 discussing the foundation of the OAS); fuller examples are found in Paul Henissart, *Wolves in the City: The Death of French Algeria* (London: Rupert Hart-Davis, 1970), esp. 39–42, 224, 364;

Alexander Harrison, *Challenging de Gaulle: The OAS and the Counter-Revolution in Algeria, 1954–1962* (New York: Praeger, 1989); and Jacques Delarue, *L'OAS contre de Gaulle* (Paris: Arthème Fayard, 1981).

56. See Georges Bidault, *Resistance,* 209–57; and another account of their adventures in Suzanne Bidault, *Souvenirs* (Caen: Ouest-France, 1987), 115–58.

57. See Raoul Girardet and Pierre Assouline, "Algérie Française," *Singulièrement Libre: Entretiens* (Paris: Perrin, 1990), ch. 8; on torture specifically, 139–43. Girardet's contemporary analysis of the French army's outlook and political function at the time of the Algerian War may be discerned through his edited volume, *La Crise militaire française, 1945–1962: Aspects sociologiques et idéologiques* (Paris: Presses de la Fondation Nationale des Sciences Politiques, 1964).

58. For the political context, see two excellent works: Emmanuel Sivan, *Communisme et nationalisme en Algérie, 1920–1962* (Paris: Presses de la Fondation Nationale des Sciences Politiques, 1976); Jean-Pierre Azéma (ed.), *Les Français et la guerre d'Algérie* (Paris: Presses de la Fondation Nationale des Sciences Politiques, 1990).

59. "Le message du chef de l'Etat à la Nation: Texte du message du général de Gaulle à la Nation, diffusé le dimanche 23 avril 1961, à 20 heures," reproduced in Vaïsse, *1961: Alger, le putsch,* 168–69.

60. See Maurice Vaïsse, *1961: Alger, le putsch,* esp. 34–37; Cf. Jean Planchais's review of this book in *Le Monde,* 16 December 1983, 26.

61. Meisel, *Fall of the Republic,* 195–96; on the attempted putsch, see also Gandy, *Salan,* 353–64; and Williams, *Wars, Plots and Scandals,* 192–203.

62. Meisel, *Fall of the Republic,* 201. For a more general and informative treatment of the theme of the French military's disorientation since the start of the Indochina war, see George Armstrong Kelly, *Lost Soldiers: The French Army and Empire in Crisis, 1947–1962* (Cambridge, Mass.: MIT Press, 1965).

63. Vaïsse, *1961: Alger, le putsch,* 134–35. Cf. Menard, *Army and the Fifth Republic,* 217–21.

64. Vaïsse, *1961: Alger, le putsch,* 135–40.

65. Ibid., 93; and testimony of Hélie de Saint-Marc at his trial in 1961 in Beccaria, *Saint-Marc,* 221–40.

66. Testimony of General Raoul Salan in *Le Procès du général Raoul Salan: Sténographie complète des audiences; Réquisitoire; Plaidoiries; Verdict* (Paris: Nouvelles Editions Latines, 1962), 84. Cf. Gandy, *Salan,* 382–403.

67. Admiral Ortuli, "Le Général de Gaulle, soldat-écrivain-homme d'Etat," *Revue de défense nationale* [hereafter *rdn*], 15 (April 1959): 584.

68. General Edmond Ruby, "L'Armée en péril," *Ecrits de Paris,* no. 196 (September 1961): 70.

69. For the French cabinet's obstruction of de Gaulle's wish to confirm the death sentence against Jouhaud, see Demaret and Plume, *Target de Gaulle,* 124–25; on the sentencing and execution of Bastien-Thiry, see ibid., 203–4, 210–16, and a sympathetic biography by a former Polytechnique classmate: Gilbert Labadie, *Bastien-Thiry, mon camarade* (Paris: Gilbert Labadie, 1989).

70. For the conversion of French officers such as Argoud (whose attitude toward the native Algerian moslems was "Protect them; Involve them; Control

them") to the employment of "psychological warfare" techniques (later embraced as "psy-ops" by the U.S. armed forces) and propaganda/ideological operations designed to win the hearts and minds of Algerian Arabs, not just physical control of the souks or the Sahara, see Vaïsse, *1961: Alger, le putsch*, 55–57, 65, 93. Meisel (*Fall of the Republic*, 186–88) discusses the related tendency of some activist officers in Algeria, such as Massu and Godard, to conclude that the army needed to assume the functions of a party, providing political leadership, a philosophy, and indoctrination alongside its conventional war-fighting and anti-terrorist functions.

71. See the account of de Gaulle's collaborator Pierre Lefranc, *Avec de Gaulle: 25 ans avec le général de Gaulle* (Paris: Plon, 1979; reprint, Paris: Presses-Pocket, 1989), esp. 147–274.

72. On this tradition of the interspersing of republican and authoritarian regimes since the 1789 Revolution, see Philip Thody, *French Caesarism from Napoleon to Charles de Gaulle* (London and Basingstoke: Macmillan, 1988).

73. Charles de Gaulle, *Mémoires d'Espoir*, vol. 1, *Le renouveau, 1958–1962* (Paris: Plon, 1970; reprint, Paris: Presses-Pocket, 1989), esp. 9–42 on the new institutions of the Fifth Republic.

74. Meisel, *Fall of the Republic*, 201.

75. Jean Planchais, "L'Armée et le tournant de 1958," *Pouvoirs: Revue française d'études constitutionnelles et politiques* (special issue: *L'Armée*) 38 (1986): 5–12 (quotation p. 10).

76. Meisel, *Fall of the Republic*, 200. Cf. Vaïsse, *1961: Alger, le putsch*, 140–41; Jacques Fauvet and Jean Planchais, *La Fronde des généraux* (Paris: B. Arthaud, 1961).

77. Address to the nation on 29 January 1960, printed in André Passeron, *De Gaulle parle, des institutions, de l'Algérie, de l'armée, des affaires étrangères* (Paris: Plon, 1962), 228. Cf. testimony of Bastien-Thiry, 2 February 1963, in *Le Procès du Petit-Clamart: Exposé des faits; Débats; Réquisitoire; Plaidoiries* (Paris: Nouvelles Editions Latines, 1963), 117. The description of Gaullist legitimacy as an "obsessional idea" is found in Soustelle, *L'Espérance*, 254; as "more legendary than historical," in Mitterrand, *Coup d'etat*, 74–75. De Gaulle's reunification of nation and regime is examined in Henry Azeau, *Révolte militaire: Alger, 22 avril 1961* (Paris: Plon, 1961), 166–72.

78. Meisel, *Fall of the Republic*, 200. Cf. Vaïsse, *1961: Alger, le putsch*, 140–41; and Hélie de Saint-Marc's account of his imprisonment at Tulle in Beccaria, *Saint-Marc*, 241–63.

79. See Planchais, "L'Armée et le tournant de 1958," esp. 9–12; also Bernard Chantebout, "La Dissuasion nucléaire et le pouvoir présidentiel," *Pouvoirs* 38 (1986): 21–32.

80. Planchais, "L'Armée et le tournant de 1958," 11.

81. See the collective work *L'Aventure de la bombe* (Paris: Plon, 1985), a volume of essays on the history of the French military atom drawn from papers presented at a colloquium convened by the Institut Charles-de-Gaulle at Arc-en-Senans, Besançon, in September 1984.

82. See Michael Dobry, "Le Jeu du consensus," *Pouvoirs* 38 (1986): 47–66.

83. Patrice Buffotot, "L'Arme nucléaire et la modernisation de l'armée française," *Pouvoirs* 38 (1986): 33–46 (esp. 39–44).

84. See the characterization at the end of Rudelle, *Mai 1958*, 293.

85. Buffotot ("L'Arme nucléaire," 45–46) fairly cautions that the dramatic alteration in the army's role imposed by de Gaulle was not achieved without "resistance" from within the officer corps and what Buffotot terms a "crisis of modernity."

86. Ibid., 45.

87. See Lawrence S. Kaplan and Kathleen A. Kellner, "Lemnitzer: Surviving the French Military Withdrawal," in Robert S. Jordan (ed.), *Generals in International Politics: NATO's Supreme Allied Commander, Europe* (Lexington: University Press of Kentucky, 1987), 93–121.

88. See Jane E. Stromseth, *The Origins of Flexible Response: NATO's Debate over Strategy in the 1960s* (New York: St. Martin's Press, 1988), 96–120 ("A Different Vision: French Responses").

89. Meisel, *Fall of the Republic*, 196. Cf. a contemporary French historical treatise on the socio-political place of the officer corps in François Kuntz, *L'Officier français dans la nation* (Paris: Charles Lavauzelle, 1960), esp. 161–84.

90. France has, of course, continued to maintain "out-of-area" basing agreements and has periodically mounted limited military interventions in the Third World, usually deploying the professional rapid-reaction regiments of the Foreign Legion from Calvi, in Corsica, and Orange, or elements of the Eleventh Parachute Division from Toulouse, along with specialist support and air force assets. For an assessment that discusses how far the French have drawn on the experiences of the colonial warfare of the 1940s and 1950s, see John Pimlott, "The French Army: From Indochina to Chad, 1946–1984," in Ian F. W. Beckett and John Pimlott (eds.), *Armed Forces and Modern Counter-Insurgency* (London and Sydney: Croom Helm, 1985), 46–76; Pierre Sergent, *2e REP: Algérie; Tchad; Djibouti; Kolwézi; Beyrouth* (Paris: Presses de la Cité, 1984). For French participation in the liberation of Kuwait in 1990–91, see Erwan Bergot and Alain Gandy, *Opération Daguet: Les Français dans la guerre du Golfe* (Paris: Presses de la Cité, 1991); Gilbert Picard, *Au Feu avec la Division Daguet* (Paris: Editions Aramon, 1991).

91. Michel Martin, *Warriors into Managers: The French military establishment since 1945* (Chapel Hill: University of North Carolina Press, 1981). Cf. Alistair Horne, *The French Army and Politics, 1870–1970* (Basingstoke and London: Macmillan, 1984), 71–92.

92. See General Georges Fricaud-Chagnaud, "L'Avenir du modèle d'armée organisée autour de la force nucléaire," *Pouvoirs* 38 (1986): 67–79; also Vaïsse, *1961: Alger, le putsch*, 92–93. The political and ideological component is stressed in earlier issues of the *Revue de défense nationale*: see Colonel Nemo, "La Guerre dans le milieu social," *rdn* 11 (May 1956): 687–700; Colonel Nemo, "La Guerre dans la foule," *rdn* 11 (June 1956): 721–34; Colonel J. Stagnaro, "Une Conception de la défense nationale," *rdn* 3 (August 1947): 147–65; Commandant Jacques Hogan, "L'Armée devant la guerre révolutionnaire," *rdn* 13 (January–February 1957): 77–89, 211–26; General André Zeller, "Armée et politique," *rdn* 13 (April 1957): 499–517; General Hublot, "Avoir L'Armée de sa politique," *rdn* 32 (November

1976): 11–28; William Coulet, "Armée, nation et discipline," *rdn* 26 (March–April 1970): 433–41, 610–24; and above all, Michel Debré's eulogy of de Gaulle, "In Memoriam," *rdn* 26 (December 1970): 1763–65.

93. Menard, *Army and the Fifth Republic,* 222–34.

Chapter 6

1. I intend to deal exclusively with Israel and Egypt, Egypt being Israel's chief Arab antagonist. The 1967 Arab strategy was essentially Egyptian, the 1973 strategy exclusively so. In either case, the 1973 Egyptian-Syrian coalition was significant only in buttressing the Egyptian strategy.

2. For a more detailed analysis, see Amos Perlmutter, "Israel's Fourth War, October 1973, Political and Military Misperceptions," *orbis* 19 (Summer 1975): 434–60.

3. Perlmutter, "Israel's Fourth War," 440.

4. For the most recent and comprehensive analysis, see Yaacov Bar-Siman Tov, "The Bar-Lev Line Revisited," *Journal of Strategic Studies* 11, no. 2 (June 1988): 149–76. See also Moshe Dayan, *Milestones* (Tel Aviv: Idanim, 1976, in Hebrew) and Yitzhak Rabin, *Service Record* (Tel Aviv: Maariv, 1979, in Hebrew).

5. Bar Siman-Tov, "Bar-Lev Line Revisited," 149 (emphasis mine).

6. Ibid., 149–50.

7. Ibid. In the years 1967, 1969, 1973, and 1974, I met for long periods to discuss IDF strategy with Generals Sharon, Adan, Tamir, Gazit, Yariv, Zeira, Dayan, Rabin, and Elazar.

8. Many books and monographs have been written about Nasser. For our purposes, relating to his role in 1967 and in the War of Attrition, the principal source is P. J. Vatikiotis, *Nasser and His Generation* (London: Croom Holm, 1978), 249–62; as well as Hrair Dekmejian, *Egypt under Nasser* (Albany: State University of New York Press, 1971); Amos Perlmutter, *Egypt, the Praetorian State* (New Brunswick, N.J.: Transaction Books, 1954); Malcolm Kerr, *The Arab Cold War, 1958–1964* (London: Oxford University Press, 1965).

9. Vatikiotis, *Nasser,* 225.

10. Ibid.

11. Ibid., 253.

12. Ibid., 255.

13. The best study of the War of Attrition and 1973 as seen by an Egyptian political figure is by Mohammed Heikal, *The Road to Ramadan: The Inside Story of How the Arabs Prepared for and Almost Won the October War of 1973* (London: Collins, 1975). See also Avi Shai's superb analysis of the Egyptian political-military preparations for the Yom Kippur War based on Egyptian war plans and intentions captured by Israel during the 1973 campaign. Avi Shai, "Egypt: Before the Yom Kippur War, the Goals of the War and the Plan of Attack," *Maarachot* 250 (March 1976): 10–21.

14. Quoted in Shai, "Egypt," 16.

15. Quoted in ibid., 17.

16. See ibid., 17–18, for the use of documents captured and analyzed by the author and myself.

17. Heikal, *Road to Ramadan,* 15.

18. Zeev Schiff, *The October Earthquake* (Tel Aviv: Zmora, 1974, in Hebrew).

19. For the diplomacy of the war, an important ingredient but one I will forego, see Nadav Safran, *Israel, the Embattled Ally* (Cambridge, Mass.: Harvard University Press, 1978), 278–316.

20. I am indebted here to Bar-Siman Tov, "Bar-Lev Line Revisited." For a critical analysis of Israeli wars including 1973, see also Emannuel Wald, *The Curse of the Broken Structures* (Tel Aviv: Idanim, 1980), esp. 97–121. Yaacov Hisddai, "The Yom Kippur War—Surprise? Victory?" *Maarachot* 246 (August 1975): 7–13.

21. Bar-Siman Tov, "Bar-Lev Line Revisited," 153.

22. Ibid.

23. Ibid., 161.

24. Ibid., 162. See also Bartov's *Dado: 48 Years Plus 20 Days* (Tel Aviv: Maariv, 1978), 1: 198–201.

25. Bar-Sian Tov, "Bar-Lev Line Revisited," 163.

26. Ibid.

27. For Wald's criticism, see *Curse of the Broken Structures,* 1–2.

28. These are also the conclusions drawn by Wald, Bar-Siman Tov, Hisdai, and myself.

29. Wald, *Curse of the Broken Structures,* 104.

30. Ibid., 113.

31. Bar-Siman Tov, "Bar-Lev Line Revisited," 164.

32. Ibid., 113.

33. *Commission of Inquiry—The Yom Kippur War—Partial Report* (Tel Aviv: Am Oved, 1975, in Hebrew).

34. Bar-Siman Tov, "Bar-Lev Line Revisited," 166.

35. Ibid., 171.

36. In regard to the surprise element of the war, see Richard Betts, *Surprise Attack* (Washington, D.C.: Brookings Institution, 1982), 68–80. Although Egypt's strategic surprise accounted for its crossing of the canal, I do not consider this a serious Israeli failure. What was serious was the underestimation of the enemy's plans and preparations. On Israel's intelligence failures, see the following: for an exhaustive bibliography, Michael Handel, *Perception, Deception, and Surprise: The Case of the Yom Kippur War* (Jerusalem: The Hebrew University Press, 1976); Avi Shalim, "Failure in National Intelligence Estimates: The Case of the Yom Kippur War," *World Politics* (April 1976): 558–59; Janis Gross Stein, " 'Intelligence' and 'Stupidity' Reconsidered: Estimation and Decision in Israel, 1973," *Journal of Strategic Studies* 3 (September 1980): 149–63. See also the excellent Lt. Col. Z. Offer and A. Kober (eds.), *Intelligence and National Security* (Israel: Ministry of Defense, 1987), in Hebrew. (Also published in *Maarachot.*)

37. See Amos Perlmutter, "Crossing the Canal of Shame," *Maariv,* 15 December 1977, pp. 1–5.

38. Aharon Zeevi, "The Egyptian Plan of Deception," in Offer and Kober (eds.), *Intelligence,* 431–38.

39. On Kissinger's diplomatic management of the Arab-Israeli conflict of 1973, see Kissinger's own memoirs.

40. Yehuda Ben-Meir, *The National Security Decision-Making: The Israeli Case* (Boulder, Colo.: Westview Press, 1986), 67.

41. Ibid., 71.

42. Ibid., 84.

43. See Zeev Schiff and Ehud Yaari, *Israel's Lebanon War* (New York: Simon and Schuster, 1984), 291–95.

Chapter 7

1. See, for example, Graham T. Allison, *Essence of Decision* (Boston: Little, Brown and Co., 1971); Herbert Simon, *Models of Man* (New York: John Wiley and Sons, 1957); John Steinbruner, *The Cybernetic Theory of Decision* (Princeton, N.J.: Princeton University Press, 1974); and Morton Halperin, *Bureaucratic Politics and Foreign Policy* (Washington, D.C.: Brookings Institution Press, 1974).

2. For an excellent overview of post–Second World War insurgencies, see Robert B. Asprey, *War in the Shadows* (Garden City, N.Y.: Doubleday, 1975). For more specific analyses, see Col. Robert N. Ginsburgh, "Damn the Insurrectos," *Military Review* 44 (January 1964): 58–70; Maj. Gerald H. Early, *The United States Army in the Philippine Insurrection* (Ft. Leavenworth, Kans.: U.S. Army Command and General Staff College, no date); and Alvin H. Scaff, *The Philippine Answer to Communism* (Stanford, Calif.: Stanford University Press, 1955).

3. Lieutenant General Samuel Williams, "MAAG-TERM Activities Nov. 55–Nov. 56" (Carlisle Barracks, Penn.: Military History Institute (MHI), November 1956), 5–7; General Cao Van Vien, et al., *The U.S. Advisor*, Indochina Monographs (Washington, D.C.: U.S. Army Center for Military History, 1980), 27, 58; Interview with Major General Ruggles by CMH, Washington, D.C., 27 February 1980; and author's interview with General Maxwell Taylor, Washington, D.C., 17 June 1982.

4. Interview with Ruggles, 27 February 1980.

5. For a detailed discussion of the principles of insurgency warfare, see David Galula, *Counterinsurgency Warfare* (New York: Praeger, 1964); Robert G. K. Thompson, *No Exit from Vietnam* (New York: David McKay Company, 1969); Douglas S. Blaufarb, *The Counterinsurgency Era* (New York: Free Press, 1977). For a specific discussion of People's War, see General Vo-Nguyen Giap, *People's War, People's Army* (New York: Praeger, 1962); and Douglas Pike, *People's Army of North Vietnam* (Novoto, Calif.: Presidio Press, 1986).

6. For a detailed discussion of Kennedy's efforts to have the army develop counterinsurgency capability and the army's reaction, see Andrew F. Krepinevich, Jr., *The Army and Vietnam* (Baltimore, Md.: Johns Hopkins University Press, 1986), ch. 2.

7. Ibid., 36–37.

8. Ibid., 50–52.

9. Combat Developments Command, "CDC Program for Analysis and Development of Counterinsurgency Doctrine," CMH, 26 February 1965.

10. Krepinevich, *Army and Vietnam*, 41–42.

11. Headquarters, U.S. Army Special Forces, Vietnam, Letter of Instructions, Number 1, "The Special Forces Counterinsurgency Program," CMH, 1 January 1965; Office of the Deputy Chief of Staff, Operations, Information Brief, "Roles and Missions, U.S. Special Forces, Republic of Vietnam," CMH, 24 February 1965; and Army Concept Team in Vietnam, Final Report, "Employment of a Special Forces Group," CMH, 10 June 1966.

12. Krepinevich, *Army and Vietnam*, 113–15.

13. See Department of the Army, Combat Developments Command, *Army Airmobility Evaluation* (Fort Benning, Ga.: January 1965); and author's interview with Lieutenant General Harry W. O. Kinnard, Washington D.C., 21 June 1982.

14. Interview with General Robert R. Williams by Colonel Ralph J. Powell and Lieutenant Colonel Phillip E. Courts, MHI, 29 March 1978.

15. Cable, Commander, U.S. Military Assistance Command, Vietnam (COMUSMACV) to Commander-in-Chief, Pacific (CINCPAC), "U.S. Troop Deployment to South Vietnam," 070340Z, CMH, June 1965, 1–3. The total comprised thirty-four U.S. and ten allied (South Korean and Australian) battalions.

16. General William C. Westmoreland, *A Soldier Reports* (Garden City, N.Y.: Doubleday, 1976), 145.

17. *The Pentagon Papers*, Senator Gravel edition (Boston: Beacon Press, 1971), 4: 291–94; and General William B. Rosson, "Four Periods of American Involvement in Vietnam: Development and Implementation of Policy, Strategy, and Programs, Described and Analyzed on the Basis of Service Experience at Progressively Senior Levels" (Ph.D. diss., University of Oxford, 1978), 204.

18. Office of the Secretary of Defense, Office of Systems Analysis, "Southeast Asia Analysis Report," April 1967, 7.

19. Admiral U. S. Grant Sharp and General William C. Westmoreland, *Report on the War in Vietnam* (Washington D.C.: U.S. Government Printing Office, 1968), 137.

20. Office of Systems Analysis, Memo for Secretary of Defense, "Force Levels and Enemy Attrition," 4 May 1967, quoted in *Pentagon Papers*, 4: 461.

21. See Westmoreland, *A Soldier Reports*, 227; *Pentagon Papers*, 4, 442; and Sharp and Westmoreland, *Report on the War in Vietnam*, 134–35.

22. Krepinevich, *Army and Vietnam*, 180.

23. Ibid., 250.

24. For an excellent analysis of the events surrounding this review of U.S. policy in Vietnam, see Herbert Y. Schandler, *The Unmaking of a President* (Princeton, N.J.: Princeton University Press, 1977).

25. See Major Robert A. Doughty and Major Robert V. Smith, "The Command and General Staff College in Transition, 1946–1976" (Ft. Leavenworth, Kans.: U.S. Army Command and General Staff College, no date), 56; and Donald B. Vought, "Preparing for the Wrong War," *Military Review* 57 (May 1977): 16–34.

26. See Colonel Harry G. Summers, *On Strategy: The Vietnam War in Context* (Carlisle Barracks, Penn.: U.S. Army War College, April 1981). See also General Bruce Palmer, *The Twenty-Five Year War* (Lexington: University of Kentucky Press, 1984).

27. Summers, *On Strategy,* 15, 46–47, 56, 60, 70, 76. For a brief critique of Summers' Laotian incursion strategic alternative, see Krepinevich, *Army and Vietnam,* 262–64.

28. Richard Halloran, "U.S. Will Not Drift into Combat Role, Weinberger Says," *New York Times,* 29 November 1984. The six tests state that the U.S. combat forces should not be committed to an intervention unless: U.S. vital interests are at stake; the forces are committed in sufficient numbers and with sufficient support to win; the forces are provided with clearly defined political and military objectives; the relationship between the forces and objectives is continually reassessed and adjusted as necessary; Congress and the American people support the commitment; and all other efforts to resolve the problem have been exhausted. See Caspar W. Weinberger, *FY 1987 Annual Report to the Congress* (Washington D.C.: U.S. Government Printing Office, 1986), 78–79.

29. Ronald W. Reagan, *National Security Strategy of the United States* (Washington, D.C.: U.S. Government Printing Office, January 1988), 8, 31, 34–35.

30. Department of the Army, Field Manual 100–5, *Operations,* 5 May 1986, 6.

31. See Department of the Army, Field Manual 90–4, *Air Assault Operations,* March 1987; Department of the Army, Field Manual 7–30, *Infantry, Airborne, and Air Assault Brigade Operations,* 24 April 1981; and Department of the Army, Field Manual 31–22, *Command, Control, and Support of Special Forces Operations,* 23 December 1981.

32. See U.S. Army Command and General Staff College, Field Circular 71–101, *Light Infantry Division Operations,* 22 June 1984.

33. U.S. Army Command and General Staff College, Circular 351–1, *U.S. Army Command and General Staff College Catalog,* Academic Year 1985–86, 22 May 1985.

34. Michael Massing, "The Military: Conventional Warfare," *Atlantic Monthly* 265 (January 1990): 32.

35. See Richard Halloran, "82nd Airborne Trains for Central American Combat," *New York Times,* 3 November 1986, 3; Julia Preston, "U.S. Latin War Games Aim for Familiarity," *Washington Post,* 31 January 1987, 9; Richard Halloran, "700 U.S. Paratroopers Fighting Mock Battles in Hills of Honduras," *New York Times,* 10 February 1987; Glenn Garvin, "U.S. Bares its Claws in Honduras Exercise," *Washington Times,* 12 May 1987, 1; and Peter Ford, "U.S. Wargames Leave Honduras Unmoved—But Not Nicaragua," *Christian Science Monitor,* 13 May 1987, 8.

36. Lieutenant Colonels A. J. Bacevich, James D. Hallums, Richard H. White, and Thomas F. Young, "American Military Policy in Small Wars: The Case of El Salvador" (Unpublished paper, Institute for Foreign Policy Analysis, March 1988).

37. For a general discussion of pacification activities in Vietnam prior to Tet-

68, see Robert W. Komer, *Impact of Pacification on Insurgency in South Vietnam* P-4443 (Santa Monica, Calif.: Rand Corporation, August 1970).

38. See Krepinevich, *Army and Vietnam*, 65–90.

39. Cited in William Manchester, *One Brief Shining Moment* (Boston: Little, Brown and Company, 1983), 224–25; and Stanley Karnow, *Vietnam: A History* (New York: Viking Press, 1983), 395.

Chapter 8

1. This idea, and indeed much of the background for this case, is drawn from Gerald Segal, *Defending China* (Oxford: Oxford University Press, 1985).

2. Segal, *Defending China*. Also see Allen Whiting, *The Chinese Calculus of Deterrence* (Ann Arbor: University of Michigan Press, 1975), and Melvin Gurtov and Byung-Moo Hwang, *China under Threat* (Baltimore, Md.: Johns Hopkins University Press, 1980).

3. See also Robert Ross, *The China Tangle* (New York: Columbia University Press, 1988), and King C. Chen, *China's War with Vietnam, 1979* (Stanford, Calif.: Hoover Institution Press, 1987).

4. Harlan Jencks, "China's Punitive War on Vietnam," *Asian Survey* 8 (1979).

5. David Goodman, *Deng Xiaoping* (London: Penguin, 1990).

6. Ellis Joffe, *The Chinese Army after Mao* (London: Weidenfeld, 1987). See also Harlan Jencks, *From Muskets to Missiles* (Boulder, Colo.: Westview Press, 1982).

7. For a general analysis of this period, see Harry Harding, *China's Second Revolution* (Washington, D.C.: Brookings Institution Press, 1988).

8. See, generally, Michael Yahuda, *Chinese Foreign Policy after Mao* (London: Macmillan, 1983).

9. Gerald Segal, *Sino-Soviet Relations after Mao*, Adelphi Papers No. 202 (London: International Institute for Strategic Studies, 1985).

10. Joffe, *Chinese Army*, and also Gerald Segal, "The PLA under Modern Conditions," *Survival* (July–August 1985): 146–57.

11. See various chapters in Gerald Segal and William Tow (eds.), *Chinese Defence Policy* (London: Macmillan, 1984); Gary Klintworth, *China's Modernization* (Canberra: Australia Government Printing Service, 1989); and Rosita Dellios, *Modern Chinese Defence Strategy* (London: Macmillan, 1989).

12. Yitzhak Shichor, "Unfolded Arms: Beijing's Recent Military Sales Offensive," *Pacific Review* 1, no. 3 (1988): 320–30 and, generally, Anne Gilks and Gerald Segal, *China and the Arms Trade* (London: Croom Helm, 1985).

13. Gerald Segal, "As China Grows Strong," *International Affairs* 64 (Summer 1988): 351–65.

14. Yitzhak Shichor, "Defence Policy Reform," in Gerald Segal (ed.), *Chinese Politics and Foreign Policy Reform* (London: Kegan Paul International for the Royal Institute for International Affairs, 1990).

15. Simon Long, "Political Reform," in Barbara Krug, Simon Long, and Gerald Segal, *China in Crisis* (London: Royal Institute for International Affairs

Discussion Paper, 1989), and several chapters in Gerald Segal and Akihiko Tanaka (eds.), *Chinese Reforms in Crisis* (London: Royal Institute for International Affairs, 1989).

16. Well noted in Harlan Jencks, "The Chinese PLA, 1949–1989," in David Goodman and Gerald Segal (eds.), *China at Forty* (Oxford: Oxford University Press, 1989). See also Michael Swaine, *The Military and Political Succession in China* (Santa Monica, Calif.: Rand Corporation, 1992); Eberhard Sandschneider, "Military and Politics in the PRC," in June T. Dreyer (ed.), *Chinese Defense and Foreign Policy* (New York: Paragon, 1988).

17. Michael Yahuda, "Sino-American Relations," in Segal (ed.), *Chinese Politics;* A. James Gregor, *Arming the Dragon* (Washington, D.C.: University Press of America, 1987); and Harvey Nelsen, *Power and Insecurity* (Boulder, Colo.: Lynne Reinner, 1989).

18. Gilbert Rozman, *The Chinese Debates about Soviet Socialism* (London: I. B. Tauris, 1987), and also Segal, *Sino-Soviet Relations.*

Contributors

Martin S. Alexander

Alexander was educated at St. Antony's College, Oxford, and at the Université de Paris IV Sorbonne before holding a Franco-British Council Research Fellowship and then a lectureship at the University of Southampton. In 1988–89 he was a John M. Olin Fellow at Yale University, and in 1991–92 a visiting associate professor at the United States Naval War College. Alexander is author of *The Republic in Danger: Maurice Gamelin and the Politics of French Defence* and co-editor of *The French and Spanish Popular Fronts.*

George J. Andreopoulos

Andreopoulos studied history, law, and international relations at the universities of Chicago and Cambridge. He has written on diplomatic history, international security, and international human rights. His forthcoming books include *The Laws of War: Constraints on Warfare in the Western World* (with Michael Howard and Mark Shulman) and *Genocide: The Conceptual and Historical Dimensions.* He is currently lecturer at Yale University. From 1989 to 1993 he served as associate director of the Orville H. Schell Center for International Human Rights at Yale University.

Philip C. F. Bankwitz

Bankwitz served with the French 2e Division Blindée of General Leclerc from 1944–45 and was awarded the French Croix de Guerre for the period 1939–45. Educated at Harvard University, Bankwitz has held fellowships from the Guggenheim Foundation, the American Council of Learned Societies, and the Fondation Camargo. In 1991 Bankwitz retired from Trinity College, Hartford, where he was distinguished professor of French history. He has written *Maxime Weygand and Civil-Military Relations in Modern France* and *Alsatian Autonomist Leaders, 1919–1947*, as well as articles that have appeared in the *Journal of Modern History* and *French Historical Studies*.

John Gooch

Gooch is professor of international history at the University of Leeds. He was educated at King's College, University of London, where he took a First in history and a doctorate in war studies. He was appointed the founding secretary of the Navy Senior Research Fellow at the United States Naval War College in 1985–86, where he was a member of the strategy department and taught in the College of Naval Warfare. In 1988 he was visiting professor of military history at Yale University.

Gooch is co-editor of the *Journal of Strategic Studies* and the chairman of the Army Records Society. He is also a vice-president of the Royal Historical Society. He has published ten books on aspects of British and European military history. His most recent works are *Decisive Campaigns of the Second World War* and *Military Misfortunes: The Anatomy of Failure in War*.

R. J. B. Knight

Knight is the chief curator at the National Maritime Museum, Greenwich, England. While custodian of manuscripts there, he edited the two-volume *Guide to Manuscripts* (1977, 1980). His early research into the British dockyards during the American Revolution was followed by his exploration of the administrative and technological problems of the late eighteenth-century British navy. Most recently he was the coordinating editor for the eighteenth-century portion of the Navy Records Society centenary volume, *British Naval Documents, 1204–1960*.

Andrew F. Krepinevich, Jr.

Krepinevich is the director of the Defense Budget Project. He has served as assistant to the director for net assessment in the Office of the Secretary of Defense and as assistant for special projects for three secretaries of defense. Krepinevich is a graduate of West Point, and he holds

M.A. and Ph.D. degrees from Harvard University. He is a member of the adjunct faculty at George Mason University and at the Nitze School of Advanced International Studies.

Peter Paret

Paret is professor of history at the Institute for Advanced Study at Princeton University. Among his works related to the subject of this volume are *Yorck and the Era of Prussian Reform, Clausewitz and the State,* and *Understanding War.* In 1993 he received the Thomas Jefferson Medal of the American Philosophical Society for distinguished achievements in the humanities.

Amos Perlmutter

Perlmutter is a professor of political science at American University. He is the author of twelve books in the field of comparative and international politics. His most recent book is *FDR and Stalin: A Not So Grand Alliance, 1943–1945.* He is the editor of the *Journal of Strategic Studies* and *Security Studies.*

Gerald Segal

Segal is a senior fellow at the International Institute for Strategic Studies, London, and editor of the *Pacific Review.* His recent, related publications include *Rethinking the Pacific, The Fate of Hong Kong,* and *The World Affairs Companion.* He is also editor of *Chinese Politics and Foreign Policy Reform* and co-editor of *China in the Nineties.*

Harold E. Selesky

Selesky is associate professor of history at the University of Alabama, Tuscaloosa, and director of the Master of Arts in Military History Degree Program at the Air University, Maxwell Air Force Base, Montgomery. He received his doctorate from Yale University in 1984 and was an instructor in history and dean of Timothy Dwight College at Yale before moving to Alabama in 1991. His publications include *War and Society in Colonial Connecticut.*

Brian R. Sullivan

Sullivan received his education at Columbia University. After serving in Vietnam, where he won a Silver Star and a Purple Heart as an officer in the Marine Corps, he taught history at Yale University and the United States Naval War College. He has written numerous articles on Italian military history and is co-author of *Il Duce's Other Woman*. Sullivan is a senior fellow at the National Defense University.

Index

Sullivan, Brian R., 6, 59–78
Summers, Harry, 135
Supply system: British defeat in Boer
  War and, 49; Italian military system
  and, 63, 64, 65, 70, 72; Prussian
  military system and, 34
Susini, Jean-Jacques, 93
Syrett, David, 11
Syria, 178n1

Tactics: British military system and, 49,
  51–53, 58; Chinese defeat and, 146–
  47; Israeli strategic doctrine and,
  109, 117; Italian military system and,
  67, 72–73; Prussian military system
  and, 34–35
Tal, Israel, 103, 115, 116, 120
Tamir, Avraham, 122
Taylor, Maxwell, 128
Technological innovation: British
  munitions and, 49–51; British shipbuilding and, 12, 21–23; Egyptian
  forces in Israel and, 117; French
  nuclear force and, 100–101. See also
  Shipbuilding
Terrain, 63–64, 66, 87
Tet Offensive (Vietnam), 133
Third World conflicts: France and,
  177n90; Military Assistance Advisory Groups and, 125–26; Six Tests
  for commitment of U.S. forces to,
  135–36, 141, 182n28; U.S. mili-
  tary performance and, 139–41; U.S.
  military preparation and, 136–39
Three Mountains offensive, 70–71
Tint, Herbert, 86, 87, 88
Tixier-Vignancourt, Jean-Louis, 93
Tournoux, Jean-Raymond, 101
Tov, Bar-Siman, 108, 116, 117
Tracy, Nicholas, 11
Training: British military system and,
  44–45, 49, 51–53, 54–57; Italian
  military system and, 63, 64, 65; Prus-
  sian military system and, 34, 35; U.S.
  Army and, 129–30, 138, 141
Transport. See Supply system
Treves, Claudio, 66
Truman, Harry S. (U.S. president), 84

United States: boundaries of defeat-
  recovery cycle and, 8; China and,

149, 155; policy review in, 133, 134;
  Six Tests and, 135–36, 141, 182n28.
  See also American Revolutionary War,
  British defeat in; U.S. Army
U.S. Army: after Vietnam, 134–41;
  airmobile forces in, 130–31, 134;
  attrition strategy and, 132–33, 134;
  conventional warfare orientation and,
  125–39; counterinsurgency doctrine
  and, 8, 128–30, 134, 135, 136–38;
  intervention in El Salvador and, 139–
  41; "low-intensity conflicts" and,
  136–37, 138; manuals, 134, 137–38;
  Special Forces (Green Berets), 130

Vaïsse, Maurice, 89, 95
Vatikiotis, P. J., 109–10
Vaudrey, Roland, 92
Vergennes, Charles, 21
Viet Cong, and People's War strategy,
  125–27
Vietnam. See China's defeat in 1979
  war; Indochina, French war in;
  U.S. Army
Vittorio Emanuele II (king of Italy), 61
Vittorio Emanuele III (king of Italy),
  62, 69

Wald, Emanuel, 116
War of Attrition (1968–1970), 103, 109,
  110–14, 115, 119
War Office (Great Britain), 42–43,
  45–47
Weinberger, Caspar, 135, 136, 141, 142
Westmoreland, William, 131, 132,
  133, 135
Weygand, Maxime, 79
Wheeler, Earle, 128
Williams, Robert, 130
Williams, Samuel ("Hanging Sam"),
  126
Williams, Thomas, 23
Wilson, Guy Fleetwood, 46
Wolseley, Garnet, 40, 43, 47, 49
World War I, 57, 59
Wyndham, George, 40

Yariv, Aharon, 103, 122
Yom Kippur War, 113–18